Quarantine Treatments for Pests of Food Plants

Studies in Insect Biology

Michael D. Breed, Series Editor

Quarantine Treatments for Pests of Food Plants, edited by Jennifer L. Sharp and Guy J. Hallman

Nourishment and Evolution in Insect Societies, edited by James H. Hunt and Christine A. Nalepa

Advances in Insect Rearing for Research and Pest Management, edited by Thomas E. Anderson and Norman C. Leppla

Diversity in the Genus Apis, edited by Deborah Roan Smith

Quarantine Treatments for Pests of Food Plants

EDITED BY

Jennifer L. Sharp
and Guy J. Hallman

Routledge
Taylor & Francis Group

LONDON AND NEW YORK

First published 1994 by Westview Press

Published 2019 by Routledge
52 Vanderbilt Avenue, New York, NY 10017
2 Park Square, Milton Park, Abingdon, Oxon OX14 4RN

Routledge is an imprint of the Taylor & Francis Group, an informa business

Library of Congress Cataloging-in-Publication Data
Sharp, Jennifer L.
    Quarantine treatments for pests of food plants / edited by Jennifer L. Sharp and Guy J. Hallman.
        p. cm. —(Westview studies in insect biology)
    Includes bibliographical references and index.
    1. Plant quarantine. 2. Food crops—Disease and pests—Control.
I. Hallman, Guy J. II. Title. III. Series.
SB980.548 1994
632'.93—dc20   93-50700
                                                                        CIP

ISBN 13: 978-0-367-28490-9 (hbk)
ISBN 13: 978-0-367-30036-4 (pbk)

Contents

Preface ix

1 APHIS, *Michael J. Shannon* 1

2 Predicting the Establishment of Exotic Pests in Relation 11
to Climate, *Susan P. Worner*

3 Statistical Methods for Quarantine Treatment Data 33
Analysis, *Victor Chew*

4 Statistical Analyses to Estimate Efficacy of Disinfestation 47
Treatments, *Jacqueline L. Robertson, Haiganoush K. Preisler,
E. Ruth Frampton, and John W. Armstrong*

5 Fumigation, *Victoria Y. Yokoyama* 67

6 Pesticide Quarantine Treatments, *Neil W. Heather* 89

7 Irradiation, *Arthur K. Burditt, Jr.* 101

8 Cold Storage, *Walter P. Gould* 119

9 Hot Water Immersion, *Jennifer L. Sharp* 133

10 Heated Air Treatments, *Guy J. Hallman and John W. Armstrong* 149

11 Radio Frequency Heat Treatments, *Guy J. Hallman and* 165
Jennifer L. Sharp

12 Controlled Atmospheres, *Alan Carpenter and Murray Potter* 171

13 Commodity Resistance to Infestation by Quarantine   199
    Pests, *John W. Armstrong*

14 Pest Free Areas, *Connie Riherd, Ru Nguyen, and James R. Brazzel* 213

15 Systems Approaches to Achieving Quarantine   225
    Security, *Eric B. Jang and Harold R. Moffitt*

16 Combination and Multiple Treatments, *Robert L. Mangan*   239
    and Jennifer L. Sharp

17 Quality and Condition Maintenance, *Roy E. McDonald*   249
    and William R. Miller

About the Contributors   279
Index   283

Preface

Horticultural industries compete to meet the demand from a growing population for fresh fruits and vegetables that look good, taste great, and are free of food pests and harmful chemicals. As new food products and larger volumes of them are moved into new areas, risk of pest introduction increases. Because consumers demand high-quality fresh fruits and vegetables, quarantine treatments must not damage product quality. Quarantine agencies must be knowledgeable of current treatments and ways to ensure the uninterrupted movement of fresh food products without jeopardizing quarantine security. This responsibility is an ongoing challenge because the use of highly effective chemicals is being restricted even as novel fresh produce is being introduced and transported into new areas.

Quarantine Treatments for Pests of Food Plants is a source of condensed information on quarantine topics. It covers the roles and interactions of the United States Department of Agriculture's Agricultural Research Service and Animal and Plant Health Inspection Service and other responsible agencies worldwide whose primary task is to guarantee the uninterrupted movement of fresh food produce without jeopardizing quarantine security. It discusses the many different treatment methods used now and perhaps in the future to ensure negligible pest risk.

This book would not have been published without the continued assistance of Sharon B. Pickard, who typed much of the book, arranged the format, proofread, made corrections, and offered improvements. Gordon Millard provided the computer expertise which proved invaluable in setting the format of the book. Wilhelmina Wasik graciously helped with proofreading and correcting.

Jennifer L. Sharp
Guy J. Hallman

1

APHIS

Michael J. Shannon

The Animal and Plant Health Inspection Service (APHIS) was established in 1972. Its mission is "To provide leadership in ensuring the health and care of animals and plants, to improve agricultural productivity and competitiveness, and to contribute to the national economy and the public health." Prior to 1972, this mission resided within the mandate of the Agricultural Research Service (ARS). In a 29 October 1971 memorandum establishing the Service, the Acting Secretary of Agriculture stated, "It will remain the responsibility of ARS to provide the necessary research support required by APHIS." Thus, research authority was retained in ARS, not delegated to APHIS. In a narrow sense, APHIS could be viewed as a key client of ARS in its reliance upon ARS for scientific results to accomplish its mission. However, from an external perspective and in a practical sense, the agencies have the same clients which perceive APHIS and ARS as the same service provider. Although organizationally separated, the agencies are partners in providing services to the same stakeholder groups. Within this partnership, ARS provides the basic science on agricultural problems and APHIS operationalizes the scientific results for application to specific problems. APHIS programs provide a conduit for introducing new technologies and scientific results into pest management systems and a forum for multidisciplinary collaboration to define and solve problems. Perhaps nowhere is the transfer of ARS scientific results to the problems of departmental clients more direct than in the area of addressing phytosanitary barriers to free import/export of commodities. Direct outcomes of these ARS scientific outputs are availability of imported commodities to the American public and the access of United States (U.S.) agricultural exports to foreign markets.

APHIS Management of Phytosanitary Issues

To provide a context for exploring in more detail cooperative work between ARS and APHIS in this area, an overview of the APHIS phytosanitary quarantine program and how the program functions is necessary. In the plant health area, APHIS provides leadership, management, and coordination of national and international activities to protect the health of U.S. agricultural resources and facilitate their movement in commerce. In regard to the management of immigrant species, Figure 1.1 depicts the functional areas where this service is delivered.

This discussion focuses on APHIS regulation of plant products to deter international and interstate movement of exotic pests, pathogens, and weeds when they affect marketability of agricultural products. APHIS regulations are implemented through a system of risk analysis and permits. Figure 1.2 illustrates these processes. Most foreign countries have similar regulatory organizations, many of which use a similar process.

Based on this process, basic conditions for importation are defined and included on the importation permit, if one is granted. For situations where significant risks need to be managed, operational systems are put in place based on the risks identified and conditions for safe importation and product movement. Results from ARS researchers are a foundation for the technical integrity of this process. Phytosanitary concerns of foreign countries which may impede entry of U.S. exports are essentially the same. In general, the burden of identifying risks and level of mitigation required and supplying the means to ensure sufficient mitigation to the exporting country lies with the importing country.

These same principles and processes discussed above apply to managing regulatory pest risk in articles moved interstate.

The International Context of
Phytosanitary Issues

Most countries have quarantine agencies with similar social objectives of managing threats from exotic pests and diseases. Historically, the establishment of phytosanitary requirements on imported products has been a matter where each country defines risks of concern on its own terms and establishes entry requirements to attain desired levels of risk reduction.

In principle, however, phytosanitary restrictions on trade must be

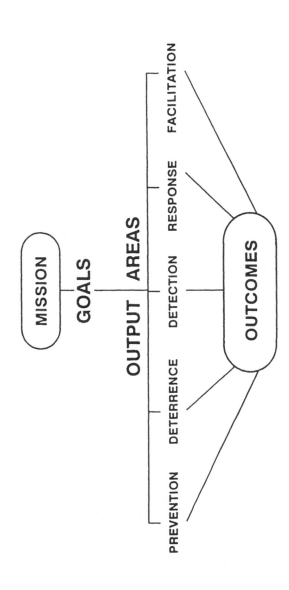

FIGURE 1.1 Management of immigrant organisms

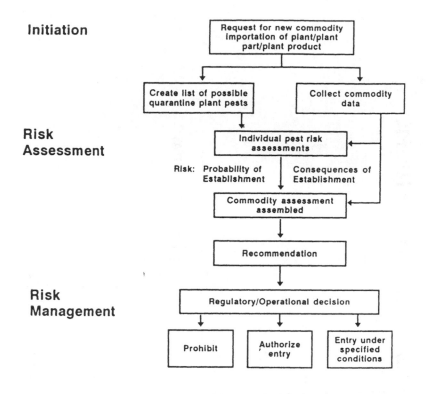

Initiation

Request for new commodity importation of plant/plant part/plant product

Create list of possible quarantine plant pests

Collect commodity data

Risk Assessment

Individual pest risk assessments

Risk: Probability of Establishment | Consequences of Establishment

Commodity assessment assembled

Recommendation

Risk Management

Regulatory/Operational decision

Prohibit

Authorize entry

Entry under specified conditions

FIGURE 1.2. APHIS system of risk analysis

scientifically justified not to constitute unwarranted barriers to free trade. The International Plant Protection Convention (IPPC) provides a framework for this type of activity. The IPPC was approved, under Article XIV of the Constitution of the Food and Agriculture Organization of the United Nations (FAO) by the Sixth Session of the FAO Conference in November 1951, and came into force on 3 April 1952 when three signatory governments ratified the Convention. Seventy-nine countries have become contracting parties to the Convention (Food and Agriculture Organization, International Plant Protection Convention 1951).

The IPPC has as its aim the strengthening of international efforts to combat important pests affecting plants and plant products and to prevent their spread across international boundaries. The Convention has 15 articles. These articles provide the establishment of official plant health organizations, uniform regulatory and quarantine procedures, and a model phytosanitary certificate with parameters for its use. These basic principles have been further defined and specified in the General

Agreement on Tariffs and Trade (General Agreement on Tariffs and Trade 1979), the Canada/U.S. Free Trade Agreement (Canada-United States Free Trade Agreement 1988), and the North American Free Trade Agreement (North American Free Trade Agreement 1993). These agreements embody basic principles for phytosanitary import/export actions as follows:

Sovereignty: Countries may use phytosanitary measures to regulate the entry of plant products.

Necessity: Institute restrictive measures only where necessary.

Minimal Impact: Adopt the least restrictive measures preventing introduction of pests.

Transparency: Publish and disseminate phytosanitary requirements.

Modification: As conditions change, modify requirements.

Harmonization: Phytosanitary measures to be based on international standards and guidelines.

Equivalence: Countries shall recognize measures that are not identical but have the same effect.

Dispute Settlement: Unresolved actions may be addressed by means of a multilateral settlement system.

The recent emphasis on quarantine principles and standards in international free trade agreements and environmental concerns over chemicals used for quarantine treatments are having significant influence on both the nature of quarantine issues to be managed and the ways they must be approached.

Operationalizing Science
to Manage Phytosanitary Risks on Commodities

The regulatory decision system utilized by APHIS (Fig. 1.3) identified key processes in identifying and managing risk which are built on foundations of sound scientific information. As illustrated by the diverse array of technologies discussed in this volume, this science can take many forms to achieve specified levels of tolerable risk. Defining these levels is a controversial area. Subjectively, agricultural production stakeholders desire no risk from the movement of agricultural products across political boundaries. Objectively, however, just as other countries accept some level of risk when importing U.S. agricultural products, some level of risk is inherent in the importation and interstate movement of plants and plant products.

The concept of zero risk is not scientifically definable. Any time a commodity is allowed entry, even with chemical or other treatment applied, there is some degree of risk. In the case of commodities subject to treatment, the risk is a function of the scientific efficacy standards established and the quality of data supporting acceptance of a given treatment or the mitigation measure; the quality of controls put in place to manage risk during harvest, handling, treatment, and shipping; and the potential for breakdowns in those controls. Many of these are not quantifiable but rather are based on subjective scientific judgment and past experience.

Baker (1939) established the probit 9 level of security as a statistical standard for development of laboratory data on quarantine treatments for fruit flies in commercial fruit. In percentages, this probit represents a mortality of 99.996832% or a survival of approximately 3.2 out of 100,000 insects. The actual security provided by this standard can vary depending on the biology of the target pest, quality controls of field treatments, and extent of insect populations within fruit.

In the case of fruit flies, acceptable tolerance levels for risk in imported fruits and vegetables are still defined the same way today with certain exceptions. Over time, additional types of exotic pests and diseases have been identified on a case-by-case basis as risk analyses are completed. Acceptable and unacceptable levels of risk and mitigation have been defined on largely a scientifically intuitive case-by-case basis.

Science is applied to phytosanitary threats associated with the movement of commodities in the following general areas:

Situational assessment of risk. Interpretation and development of scientific data relating to defining the probability of organism occurrence and colonization potential. An example of this is experimentation to show an exported item is not a host of fruit flies or specific diseases or that the organism will not colonize as a result of importing the item. Studies on transmission of fire blight on apples and the non-fruit fly host status of specific fruits and vegetables are examples of this.

Validating the efficacy of quarantine treatments for pests and commodities. Developing data through scientific protocols that proves efficacy, assesses effect on product quality, and in the case of chemicals

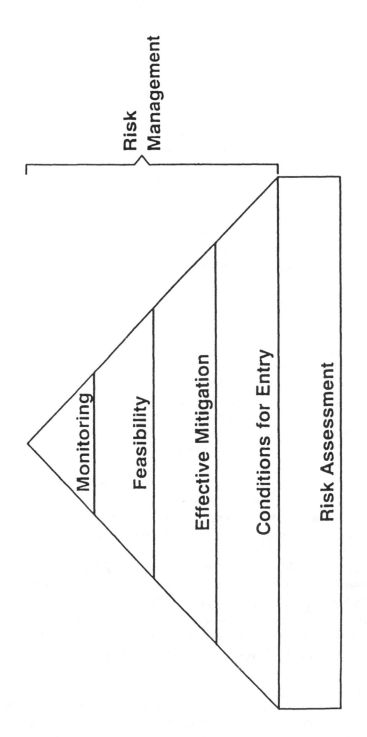

FIGURE 1.3. APHIS regulatory decision system

documents residues.   Such data are generally required on each pest/commodity combination. In some cases, especially with physical (e.g., heat) treatments, tests are required for different varieties of the same fruit.

Pest free zones. Establishment of areas for export production within a country where a quarantine object(s) does not exist, even though it is established elsewhere in that country. The technical problem here is having the technology and scientific basis to ensure a pest free area exists and can be maintained.

Under this approach to providing quarantine security, specific areas are certified. Based on survey technology and data that confirm the area is free of the organism of concern, the exporting country establishes formal, specific regulatory measures to protect the area and an ongoing detection system that ensures timely detection of any infestations in the area. These systems are dependent on scientific information, judgments about organism behavior, and survey technologies and methods.

System approaches. Implementation of multiple safeguard actions in country of export that result in a commodity which meets phytosanitary standards of the importing country.

These situational quarantine certification approaches differ from traditional quarantine commodity treatments in that the defined level of risk reduction attained is not quantifiably defined through experimental procedure to a statistically defined level (i.e., probit 9). These systems, however, must have a scientifically derived basis.

Perhaps the first example of this method of providing quarantine security was the export of Unshu oranges from Japan to the U.S. with acceptable levels of risk from citrus canker disease. In this situation, a combination of resistant varieties, buffer/host free zones, inspection, testing, and limited destinations allows for this commodity to be imported for over 20 years without undue risk from the disease. A second example is the export protocol utilized to certify Florida grapefruit exports to Japan as free of Caribbean fruit fly, *Anastrepha suspensa* (Loew). Rather than being based on a quarantine treatment providing probit 9 security, risk is mitigated by the biology of the pest in the Florida environment, trapping, bait sprays, and other safeguard actions.

As chemically based commodity treatments become less available, more reliance is being placed on such system approaches. This multiplies the number and complexity of problems to be dealt with since technical problems are defined on a production area basis rather than only a commodity basis. APHIS continues to be highly dependent on the scientific expertise of ARS scientists in the development, approval, and scientific validation of these free zone and system approaches to phytosanitary certification of import/export agricultural products.

Future Challenges

The changing nature of phytosanitary certification problems is requiring that solutions move away from adapting existing chemical/physical treatments and toward new treatment technologies such as irradiation and case-by-case system approaches. This has the effect of exponentially increasing the number of competing, researchable scientific problems. In addition, import and export issues compete for limited scientific and fiscal resources available to address these questions. Increasingly, foreign countries are becoming more demanding for assistance/advice in the conduct of research necessary to meet U.S. entry requirements. The political nature of many of these situations and the increasing influence of special interests on APHIS and ARS priorities have made their management extremely problematic. Changes occurring in the environment in which APHIS and ARS work together to deliver services to our common clients promise to pose even more complex challenges in the future.

Key among these challenges are:

1. Expanded needs for alternatives to current, broad spectrum chemical mitigation of phytosanitary problems.
2. Increased scientific complexity of risk analyses to accommodate situational factors that reduce risk.
3. Expanded demands for phytosanitary issues to be defined and solved on a cropping system rather than purely a commodity treatment basis.
4. Expanded interest and needs for international standards for treatment experimentation, pest survey, and development of technical data to address phytosanitary concerns.

Addressing these challenges requires both continuing efforts to develop and refine commodity treatments and the development and analysis of area wide pest management programs to meet phytosanitary objectives. Assessing and managing phytosanitary hazards is an increasingly visible APHIS and ARS collaborative effort that creatively applies science to produce outcomes meeting the needs of clients of the United States Department of Agriculture.

References

Baker, A. C. 1939. The basis for treatment of products where fruitflies are involved as a condition for entry into the United States. USDA Circular 551.

Canada-United States Free Trade Agreement. 1988. 27 International Legal Materials. 281.

Food and Agriculture Organization. International Plant Protection Convention. 1951. Basic Texts Volume 3, Fascicle 5, 25 pp. FAO, Rome.

General Agreement on Tariffs and Trade. 265 Basic Instruments and Selected Documents 290. 1979.

North American Free Trade Agreement. 1993. 32 I.L.M. 605.

2

Predicting the Establishment of Exotic Pests in Relation to Climate

Susan P. Worner

Establishment of an insect species in any locality is affected by environmental factors. Host availability, genetic potential of the insect, and status of competitors, diseases, predators, and parasites are the biotic factors of greatest significance. The most important abiotic factors are temperature extremes, water, and wind. Climatic factors directly affect the distribution of insect populations by limiting or enhancing rates of reproduction, growth, survival, diapause, and dispersal. Indirect effects of climate are mediated through host plant responses, natural enemies, competitors, and disease (Mochida et al. 1987). Given host availability, climatic favorability dictates if an insect species will establish in a given locality. Successful establishment subsequently depends largely on the biotic features of the environment.

Because climate exerts such a profound influence on the distribution and abundance of insects, quantification of climatic influences is of considerable interest to quarantine scientists for assessment of potential establishment and spread of exotic pests. For most of this century, knowledge of the relationships between climate and insect bionomics has been used to answer questions that concern economic entomologists: (1) where (geographically) a pest species is likely to occur, (2) when during a crop season a pest species may be expected to appear, and (3) how many pest individuals are likely to survive and reproduce (Messenger 1974). Many techniques and research results from other areas of applied entomology, particularly biological control, can be applied to the quarantine problem of potential pest establishment.

Meats (1981) suggested that given enough experimental information, the bioclimatic potential of an organism could be confidently predicted. With known cause and effect relationships between survival, growth, reproduction, and dispersal of an organism and environmental factors such as topography, soil type, resources, predators, and climatic variables, it would be easy to predict the likely establishment of a quarantine pest. However, he states that the probability of such omniscience is low.

In the 1970s, easier access to computers gave hope of new tools that could predict the biotic potential of insects. Some early techniques were eclipsed by the belief that computer simulation modeling could provide more precise answers to questions of quarantine concern. However, the application of simulation to such questions has been disappointing.

Ecological models provide a structure for the synthesis of information, data summary, and insight into complex processes. When used for prediction, an ecological model allows evaluation of possible outcomes. However, because measurements of biological parameters are inaccurate and ecological relationships are incompletely described, model output should be interpreted with caution. Thus, no absolute method exists to predict if an insect will establish in a particular locality. Quantifiable prediction techniques combined with the experience, skill, knowledge, and common sense of workers in the field must ultimately be used (Worner 1988b).

Insects and Temperature

Because of its influence on physiological systems, temperature affects every life process of an insect. Moderate temperatures affect rates of development, and temperature extremes may be lethal. At the populational level, the effect of temperature on birth rates, death rates, dispersal, and migration is a major influence on the abundance of a species. Of all abiotic environmental factors, temperature has received the most attention. When assessing the potential for establishment of a exotic pest, it is important to determine if the locality at risk will provide enough heat over the susceptible period of the potential host for all life stages of the pest to successfully complete development and reproduce (Baker 1972).

The phenology or seasonality of insect appearance and disappearance includes the timing of insect activity, development, and life cycle. Timing of active periods determines whether synchrony with the host in a potential locality is possible. To predict insect phenology it is necessary to determine the relationship between insect developmental rate and temperature. This observed relationship under controlled constant laboratory conditions tends to be nonlinear throughout the full range of temperatures that can support insect life processes. Low temperatures gradually slow development, and high temperatures inhibit development. At intermediate temperatures the development rate is almost linear. The temperature and developmental rate relationships in insects have been exploited in phenological models for the prediction of important events in insect pest life cycles to assist precise application of control measures. These same models can be used to assess if localities threatened by exotic pest introduction can meet the insect's minimum seasonal thermal requirements.

Although many developmental rate models have been proposed (Wagner et al. 1984, Higley et al. 1986), all are either linear or nonlinear. Linear models use only the more linear portion of the development rate function. The rationale for use of the linear model is that only temperatures in the linear region of the rate-temperature function are important to insect development in the field. Other models use the full nonlinear developmental rate curve. Proponents of nonlinear models argue that although the linear portion of the developmental rate curve adequately describes development for some insects, many species experience temperatures within the nonlinear regions of the developmental rate curve. Thus, use of a linear model would underestimate development at low temperatures and overestimate it at high temperatures.

Linear models use temperature summation or the degree day approach to estimate heat required for development. Intrinsic to the use of this approach is the assumption that development is proportional to temperature. Consequently, there is a lower threshold below which development does not occur and a thermal constant or constant number of degree days required to complete development. This thermal constant is derived simply as the reciprocal of the slope of the regression line fitted to developmental data obtained at constant temperatures. Similarly, the lower threshold for development is defined by (-intercept)/slope. To predict development in the field, a diurnal temperature model is usually fitted to environmental temperatures to calculate degree days (Worner 1988a). Degree days are summed daily until the required number of degree days for complete development is reached. For some species winter cold rather than lack of sufficient heat during the growing season limits distribution. Chilling degree days or total degrees of temperature below the lower threshold for development can be used to measure winter cold.

To estimate time when development is complete, nonlinear models use rate summation where developmental rate, or proportion of development predicted by a nonlinear function, is summed for each temperature at short time intervals (<1 d). Linear models may also use this method. Rate summation can be defined as:

$$D = \Sigma \ r[T(dt)]dt$$

where development D is a function of temperature T which in turn is a function of time t. Thus, the proportion of development taking place according to the temperature at each time increment is summed so that when D equals 1 the insect is considered to have completed development.

The assumptions behind the use of linear and nonlinear models for phenological prediction under natural conditions can be questioned (Worner 1992). It is difficult to determine which type of model best predicts if a location meets the thermal requirements of an exotic pest. Moreover, models that predict diurnal variation in temperature from daily minimum and maximum temperature values can generate gross error in some climates (Worner 1988a). Any developmental rate function should be used only as a guide to decide if there is enough heat during the growing season for an insect to complete its life cycle and establish in a given area. However, in comparative studies with moderately large sets of validation data, the simpler linear models perform equally well or marginally better than nonlinear models under field conditions (Stinner et al. 1988, McClain et al. 1990).

Podolsky (1984) proposed using graphs where the observed phenology or time to development of an insect from some biologically meaningful point is plotted against mean temperature. From start dates at spaced intervals over a season or year, 10 d running mean temperatures are superimposed to form a heat resources net or phenotemperature nomogram (Fig. 2.1). The likely development of an insect in different localities (or different years in a particular locality) can then be predicted graphically by plotting the corresponding running average temperatures. Podolsky (1984) gave examples using this technique for mapping potential distributions of exotic organisms. This method is attractive because it uses historical field data and, therefore, includes all factors that cannot or have not been measured. In addition, it has the advantage that information regarding periods of environmental stress, such as frost susceptible periods, can be incorporated. However, this method has a theoretical weakness. When developmental rates are observed under temperatures that fluctuate around nonlinear portions of the developmental rate relationship, different rates will be observed for different temperature regimes (Ratte 1985, Worner 1992).

Other methods that use field data to predict insect phenology are

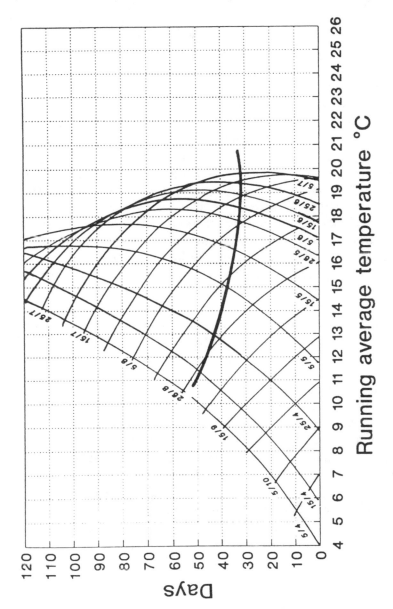

FIGURE 2.1. A phenotemperature nomogram. Concave lines are 10 d running average temperatures from selected start dates. The thick solid line represents the line of best fit (using Podolsky's method) to observed insect development in relation to temperature (Podolsky 1984).

based on degree day summations. Arnold (1959) proposed using least variability methods to determine effective lower thresholds and mean degree days required for plant development from field observations at different times or at different locations. This method has been readily applied to the prediction of insect development. Ring et al. (1983) extended this approach to include determination of effective start dates for degree day accumulation in locations where the insect normally occurs. Deciding which start date to use for degree day summation in locations where an insect does not normally occur is difficult because a start date in one locality may not be appropriate in another. Because plants accumulate the effects of prior temperature conditions, I suggest that a start date associated with some aspect the host plant phenology should be used.

Dallwitz & Higgins (1978) developed a computer program, DEVAR, that uses an appropriate numerical optimization algorithm that can derive a nonlinear development-temperature relationship from developmental time observations made under fluctuating temperature regimes. However, when tested with hypothetical developmental time data, DEVAR gave inadequate estimates of the instantaneous developmental rate function that generated the data (Worner 1992). It is unclear if this result was caused by the nature of the data set or the particular developmental rate function used, or the program itself.

Whatever model is used, a thorough understanding of insect response and model behavior under fluctuating temperature conditions is important. Of particular interest is the Kaufmann or rate summation effect (Ratte 1985). When rate summation is used, nonlinearities in the developmental rate temperature relationship (intrinsic to the nonlinear model, but introduced by use of lower and upper thresholds in the linear model) can give apparent unexpected predictions under fluctuating temperature regimes. When temperatures fluctuate around nonlinear regions of a developmental rate function, developmental rate for fluctuating temperatures will be different to the developmental rate when the temperature is held constant at the mean of the fluctuations (Kaufmann 1932, Pradhan 1945, Tanigoshi et al. 1976). The easiest way to understand the implications of model behavior for quarantine decisions is to study its performance with hypothetical data (Worner 1992).

Climate Matching

There are other aspects of climate besides temperature that influence insect distribution and abundance. For many years climographs or climatographs were used to compare combinations of climatic factors to help predict areas suitable for establishment of a particular species (Messenger 1976).

A climatograph is a graphical model that compares a combination of levels of climatic factors at localities where the target species presently occurs, with climatic levels of localities where the species may pose a threat. The simplest example is to plot the profiles of mean monthly temperatures and rainfall totals from different localities on the same graph. These can then be compared in shape and height. Similar climates will produce roughly similar graphs. A related technique is to use climatic classifications or indices to match climates. Köppen (1931) introduced a set of climatic classifications based on mean temperature and precipitation that quickly became standard (Dennill & Gordon 1990).

Another example of a climatograph is to plot mean monthly temperatures against mean monthly humidities or rainfall totals (Fig. 2.2A). Points for the 12 months of the year are joined in sequence, giving a 12 sided polygon. Different climates will produce climatographs of different shapes, locations on the temperature or rainfall axes, or both (Messenger 1974). Examples using this type of climatograph for the analysis of the potential distribution of insects include studies on alfalfa weevil, *Hypera postica* (Gyllenhal) (Cook 1925), a more general study of bioclimatic zonation of insects (Cook 1929), and a study on wheat stem sawfly, *Cephus cinctus* Norton, in Canada (Seamens 1945).

Additional information can be used to produce ecoclimatographs (Messenger 1976). For example, the survival, developmental, and reproductive limits of the insect's response to the climatic factors of interest can be superimposed (Bodenheimer 1938, Allee et al. 1949, Howe 1957, Howe & Burges 1953).

The relationship of climate to the survival and development of Australian spider beetle, *Ptinus ocellus* Brown (Howe & Burges 1953) is shown in Figure 2.2B. Climatographs of mean daily temperature and humidity for each month are given for Winnipeg, Canada; Kew, England; and Kano, Nigeria. Heavy lines enclose three zones. The outer, middle, and inner zones represent slow, moderate, and rapid development, respectively. Beyond the outer zone, conditions are lethal. Winnipeg was shown to be too cold for *P. ocellus* six months of the year.

Climatographs seem to have fallen from favor, possibly overshadowed by the promise of improved prediction provided by computer simulation. These seemingly less sophisticated techniques need to be reexamined for their utility in quarantine decisions. Use of these and other techniques requires care because some species may not be limited in their distribution as a direct effect of climatic factors.

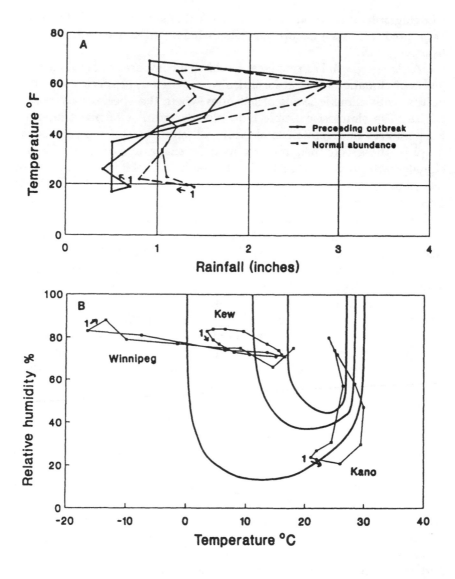

FIGURE 2.2. Examples of climatographs. A. Dashed line shows climatic conditions for each month of the year in the zone of normal abundance for army cutworm, *Euxoa auxiliaris* (Grote), in central Montana. The solid line shows climatic conditions preceding several outbreaks of the same species (Cook 1929). B. An example of an ecoclimatograph showing the relation of climate to survival and development of Australian spider beetle, *Ptinus ocellus* Brown (Howe & Burges 1953).

Messenger (1970) showed that climates outside the limits of western cherry fruit fly, *Rhagoletis indifferens* Curran, as represented by conventional climatographs, remain well within the range of values tolerated by the pest. Many similar examples can be found in biological control literature. Stiling (1990) listed many instances where temperate parasitoids have successfully established in tropical areas and vice versa. Such successful establishment in unexpected areas suggests that assessing the establishment of a quarantine threat is not as simple as matching the climate of a potential pest to that of the host location. The question is further complicated because there may be enclaves of temperateness or tropicalness in source areas so that these may not be climatically distinct from potential host locations.

Statistical Techniques

As a graphical technique, a climatograph is limited to two, and not more than three, variables to simultaneously explain the presence or absence of a species. Multivariate statistical techniques can overcome this limitation by reducing large amounts of data to a few characteristics. Principal Components Analysis (PCA) was developed to help generate hypotheses about the relationship between species composition at a site and the underlying environmental factors. Pimm & Bartell (1980) used this technique to analyze the distribution and expansion patterns of red imported fire ant, *Solenopsis invicta* Buren, in the southeastern United States (U.S.). Redundancy in a large set of correlated variables is removed by representing the variables by a new set of extracted or derived variables called factors. Because these factors explain the greater proportion of the variance data set, subsequent interpretation can be based usually on the first two or three.

For their analysis, Pimm & Bartell (1980) used a time series of mean monthly precipitation, the number of days when the temperature dropped below 0°C and the mean number of days the temperature exceeded 32°C. Three means for each of 12 months yielded 36 variables from which three were dropped because it was found that for three months of the year, there were no days in the area studied when the temperature dropped below 0°C. Representation of variables as points on a rectangular coordinate system, where the horizontal and vertical axes represent the first two factors extracted, allowed the production of a coordinate system for climatic factors called the climate space. Within this coordinate system, locations where the red imported fire ant had been observed were plotted (Fig. 2.3). This procedure translated the

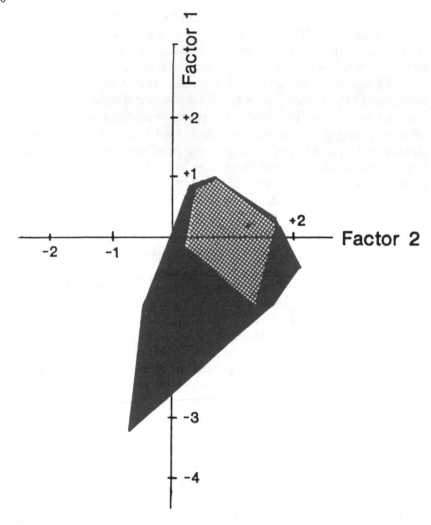

FIGURE 2.3.   Ordination of locations of red imported fire ant records: before 1965 - cross hatched, up to 1976 - shaded (Pimm & Bartell 1980).

geographic progression of the red imported fire ant's distribution over time into a progression within the climate space.   The following observations were made: from 1965 to 1976 the species increased its climate space to include a few areas that were wetter than other areas (higher Factor 2 values) (Fig. 2.3); the species spread dramatically into areas that were hotter and drier (lower Factor 1 and negative Factor 2 values), and the species showed no tendency to occupy cold dry climates (positive Factor 1 and negative Factor 2 values). Because unoccupied points in the climate space represented localities in southeastern U.S. where the red imported fire ant could potentially establish, the authors were able to cautiously predict the spread of this species.

Despite the fact that PCA is difficult to use and its interpretation complicated, it is a powerful data reduction technique. When used carefully it can increase insight into the ecological complexity of insect distribution.

Bioclimatic Indices

Different population processes have different optima and tolerance ranges in response to climatic variables. To summarize these different responses at any one time, some studies have used bioclimatic or ecoclimatic indices that quantify where an insect is likely to establish and how abundant it is likely to become. The innate capacity for increase (r_m) has been used to incorporate environmental influences on the birth rate and survival rate of a population into an index of growth potential. This population parameter is a component of the exponential growth equation:

$$N_t = N_o e^{r_m t}$$

As such, the parameter r_m describes population growth under conditions of an unlimited environment and stable age distribution. Because such conditions occur rarely in nature, the innate capacity for increase, though a biological characteristic of the insect species, gives only an indication of growth potential in a new environment.

Use of the innate capacity for increase as a bioclimatic index is based on the knowledge that temperature has a differential effect on the survival, fecundity, development rate, and longevity of insects. However, the tolerance ranges and optimum levels for these processes are not all the same (Messenger 1964a). For example, the optimum temperature for development of spotted alfalfa aphid, *Therioaphis maculata* (Buckton), is different from that for reproduction (Messenger 1964b). The population parameter r_m combines several effects of an important climatic factor such as temperature into one ecologically relevant index of environmental favorability or unfavorability related to that factor. The higher the value of r_m the higher the capacity for population increase and the more favorable the environment. For example, age specific survival and fecundity data can be obtained from insects reared in a range of temperatures. From these data, r_m values can be calculated. These r_m values provide weighted measures of rates of development, reproduction, and survival relative to temperature and, if plotted against temperature, will indicate a tolerance range for population growth. For example, Messenger (1964b) determined the effects of fluctuating and constant temperatures and humidities on the spotted alfalfa aphid and found the optimum temperature for r_m was an average fluctuating temperature of 29.4°C, the lower thermal limit was near 6°C, and the upper thermal limit was approximately 33°C.

Where density effects are minimal, it has been found that developmental period (time to first reproduction) has a greater effect on r_m than other variables (Messenger 1964a, Siddiqui & Barlow 1972). This is not surprising because temperature affects development time, and variations in time required for females to reach sexual maturity greatly affect potential population increase. Developmental period does not by itself determine r_m. A strong relationship between developmental period and the reciprocal of r_m may allow estimation of r_m under fluctuating temperature conditions when information on fecundity and survival is not known (Siddiqui & Barlow 1972, Tanigoshi et al. 1976). Other approximations to this parameter can be obtained from population growth curves at different temperatures, where the slope of the regression of log density against time estimates r_m.

Meats (1981) used the finite rate of increase (a measure of the number of times a population will multiply itself per unit time),

$$\lambda = e^{r_m}$$

to create a system of bioclimatic indices. He used an infestation index and generation index combined with isopleths (isohyets) of climatic data to reach conclusions regarding the bioclimatic potential for extension in distribution of Queensland fruit fly, *Bactrocera tryoni* (Froggatt), in Australia. All such indices should be used and interpreted with careful consideration of their limitations. Knowledge of other abiotic and biotic factors that may influence the distribution of the potential pest should be considered.

Bioclimatic Studies

Another approach to assessing potential establishment of exotic pests is to simulate realistic climatic factors in a climate or growth chamber.

Stone (1939) used a cam-activated temperature controller attached to a laboratory incubator to reproduce actual temperature patterns recorded in the field in his studies of the climatic responses of Mexican fruit fly, *Anastrepha ludens* (Loew). Messenger & Flitters (1954) also used this method to assess the bioclimatic potential of oriental fruit fly, *Bactrocera dorsalis* (Hendel), melon fly, *B. cucurbitae* (Coquillett), and Mediterranean fruit fly, *Ceratitis capitata* (Wiedemann), for various localities in the continental U.S. The growth chambers (2 by 2 meters) were air conditioned and had illuminated interiors to provide actual diurnal cycles of temperature and relative humidity. Cam-actuated control instruments reproduced the temperature and humidity patterns recorded at various field sites. This study showed that the species had a differential response to heat and cold and that favorable climates could be found only in Florida, southern Texas, and coastal southern California.

Computer Based Systems

Computer technology eliminates the tedium of numerous calculations when questions concerning insect bionomics are explored. Because many ecological relationships can be described and programmed, there is a temptation to believe that computer models and programs offer improved precision. However, because ecological measures are inaccurate, complex systems incorporating many ecological interactions tend to accumulate error (O'Neill et al. 1980, Pielou 1981). Despite these problems, computer models and computer-based ecological investigations give deeper insight into complex processes and point to gaps in our knowledge (Worner 1991).

The problem with the bioclimatic prediction methods previously discussed is combining an organism's response to numerous abiotic factors into one or two indices that are both meaningful and manageable. Sutherst & Maywald (1985) and Maywald & Sutherst (1989) described a computer based system, CLIMEX, for matching climates in ecology. The relative potential for growth and persistence of poikilotherms is compared at different times and places to determine potential distribution. CLIMEX encompasses the essence, but not the detail, of all methods discussed so far. Using monthly meteorological data, CLIMEX allows rapid worldwide comparison of climates. It can search for areas with climates similar to those within the known distribution of a species and can make predictions of likely distribution. Given further information on certain biological parameters, CLIMEX gives estimates of likely abundance. To give an idea of its simplicity, this system calculates an Ecoclimatic Index (*EI*) that describes the favorability of a location for a particular species. The ecoclimatic index is a combination of growth and stress indices that describe the response of the insect to the climate of a given location. To assess the potential for population increase, the annual mean of weekly values of a population growth index (*GI*) is determined. This growth index is reduced by four stress indices that describe the likelihood that a population will survive through a given season. The ecoclimatic index, *EI*, is defined as:

$$EI = 100 \frac{(\Sigma GI)}{52} (1\text{-}CS)(1\text{-}DS)(1\text{-}HS)(1\text{-}WS)$$

where the weekly population growth index, $GI = TI \times MI \times LI$. The indices *TI*, *MI*, and *LI* are weekly temperature, moisture and daylength indices, respectively, scaled between 0 and 1. *CS*, *DS*, *HS*, and *WS* are cold, dry, hot, and wet stress indices respectively, also scaled between 0 and 1. The stress indices describe the effects of both the duration and severity of extremely cold, dry, hot, and wet conditions.

The temperature (*TI*) and moisture (*MI*) indices are derived from relationships that involve linear interpolation between four levels. For example, the moisture index (*MI*), is shown in Figure 2.4. This index can be interpreted as a direct effect of moisture on the insect population or an indirect effect mediated by the condition of the host plant. The four parameters that describe the population response to soil moisture (*MI*) are as follows: *SM0* and *SM3* are, respectively, critical thresholds where excessive dryness and wetness of soil are lethal. At or below these levels *MI* = 0. The moisture index increases linearly from *SM0* up to an optimum level for population growth at *SM1*. Growth is optimum between the two intermediate soil moisture levels *SM1* and *SM2*, where *MI* = 1.

The light index (*LI*) incorporates an insect's response to daylength and is controlled by a single parameter setting according to the nature and the extent of the response.

The stress indices have a nonlinear cumulative effect over periods of adverse conditions. The effective stress value (*V*) for each index is defined as:

$$V = g(S) \cdot t$$

where $g(S)$ is a stress function dependent on the magnitude of the stressful condition *S*, and *t* is time in weeks during which the stressful condition is experienced. Each stress function, $g(S)$, is linear and, therefore, has two parameters: a threshold parameter that determines the level of the adverse condition at which stress starts to accumulate (e.g.,

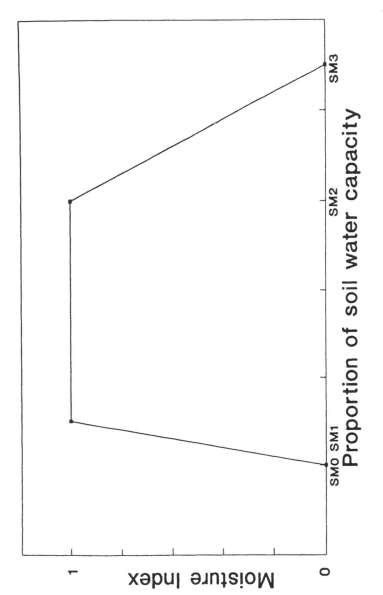

FIGURE 2.4. Moisture Index (MI) as a function of soil moisture condition. Parameters are described in text (Sutherst & Maywald 1985).

an extreme temperature) and a slope or rate parameter which determines the rate of accumulation of stress. The yearly stress index for each condition is the sum of the weekly values scaled between 0 and 1.

CLIMEX uses standard meteorological data of maximum and minimum daily temperatures, evaporation or relative humidity, and rainfall. Long term weather data can be used to compare bioclimatic favorability between locations or, using more detailed data, to compare bioclimatic favorability between years at the same location, for a target species.

By concentrating on gross features of the response of an organism to its environment, CLIMEX can provide useful information when data are scarce. However, determination of the parameters required for any investigation is difficult. For example, the response of insects to climatic stress is rarely studied. Despite this, reasonable estimates for parameters can be found.

Initial values for parameters may be determined either by a literature search, laboratory data, or empiricism. Decisions concerning the final parameters are made by iteration, i.e., the climatic response of the insect indicated by CLIMEX is compared with known responses at several well chosen localities. Differences between observed and predicted responses are then used to adjust parameters in an iterative fashion until approximate correspondence is achieved. These final parameters are then used to compare the response of the insect within climates of interest. Thus CLIMEX provides correlative measures for which statements concerning mechanisms and causation can only be speculative. Given that even thoroughly researched climatic response parameters frequently differ, together with the objective of the examination of gross features of bioclimatic potential, the process of parameter tuning or adjustment is justified. Unfortunately, it is often the only approach. Because the user has considerable flexibility, estimates of parameters must be biologically reasonable and within bounds suggested by prevailing knowledge.

CLIMEX has been applied to several quarantine problems (Worner 1988b) such as the potential spread of screwworm fly, *Cochliomyia hominivorax* (Coquerel), from Libya to other African countries (Palca 1990). It was estimated that this fly was most likely to thrive in the more tropical regions of Africa and could develop permanent populations along the Mediterranean coast. The Sahara formed a natural boundary which would prevent the fly from moving directly south, but if it reached the Nile river in neighboring Egypt it could migrate to more hospitable climates south and possibly into Europe.

CLIMEX provides graphical output where EIs can be plotted on an outline map of a particular area or country. It provides graphs of the seasonal variation in growth and temperature indices and gives accumulated degree days for each locality. CLIMEX can also be used to provide information concerning the possibility of increased quarantine threat as the result of climate change (Frampton 1990). A desirable development is the integration of CLIMEX with a geographic information system (GIS). Modern GIS incorporate advanced data management techniques and graphics to offer flexibility in the acquisition, retrieval, analysis, and display of spacial information (Forer & Chalmers 1987). Examples of GIS applied to entomological problems are the prediction of forest pest outbreaks such as spruce budworm, *Choristoneura fumiferana* (Clemens), in New Brunswick (Jordan & Vietinghoff 1987) and gypsy moth, *Lymantria dispar* (L.), in Michigan (Montgomery 1987). The use of CLIMEX or other biotic potential prediction systems in combination with a GIS would allow the use of currently available environmental data to quickly correlate insect response to a larger variety of environmental parameters.

Conclusions

None of the techniques discussed can answer all questions concerning pest establishment. CLIMEX can quickly provide much needed information for subjective risk assessment; however, it is a relatively coarse method of data summary. The best approach would be a combination of methods. Results of different methods which agree would suggest localities where predictions are more reliable. Discrepancies would indicate localities where predicted establishment is more speculative. Such analyses would be certain to identify gaps in our knowledge and allow the generation of new hypotheses concerning insect distributions.

Gilpin (1990) in a review of *Biological Invasions: A Global Perspective* (Drake et al. 1989) assessed the work of contributing authors and concluded that, given the complexity of the problem, there can never be a reliable scheme to predict the success of an invading species and the quest for true case by case predictability should be renounced. Inaccessibility of data limits real progress and more effort should be put into establishing historical computer data bases available on line to all researchers. Such data bases would allow existing predictive techniques to be properly validated and improved and new ones developed.

The decision to quarantine a foreign pest is complex as it is based on assessment of its potential impact on the biological, economic, and environmental features of a region or country. Such a decision involves the concepts of risk analysis and risk management to provide protection against entry and establishment of potentially harmful pests while facilitating international trade and free movement of people (Australian Quarantine and Inspection Service 1991). Because decision makers may have variable information or use different rules, there are no universally acceptable options (North American Plant Protection Organization 1991). For example, some countries may use quarantine as a non tariff barrier to trade and some may propose an outright no risk policy. Therefore, relevant international organizations are attempting to establish an internationally recognized standard for pest risk assessment (Australian Quarantine and Inspection Service 1991, North American Plant Protection Organization 1991). Because quarantine decisions are based on economic and environmental factors, as well as biological factors, they tend to be very conservative. Any insect that may remotely be considered a potential pest will probably be subject to quarantine despite what predictions are made by the techniques discussed here. However, ecoclimatic assessment will continue to be important for decisions concerning implementation of monitoring and eradication programs and quarantine treatments.

Acknowledgments

I thank Ruth Frampton, Bruce Chapman, and Eric Scott for their helpful comments and review of this chapter.

References

Allee, W. C., A. E. Emerson, O. Park & T. Park. 1949. *Principles of Animal Ecology*. Philadelphia: W.B. Saunders Co.

Arnold, C. Y. 1959. The determination and significance of the base temperature in a linear heat unit system. Proceedings, American Soc. Hortic. Sci. 74: 430-445.

Australian Quarantine and Inspection Service. 1991. The application of risk management in agricultural quarantine import assessment. A discussion paper. Department of Primary Industries and Energy, Canberra, Australia.

Baker, C. R. B. 1972. An approach to determining potential pest distribution. European Plant Protection Organization Bul. No. 3: 5-22.

Bodenheimer, F. S. 1938. *Problems of Animal Ecology*. Oxford: Clarendon Press.

Cook, W. C. 1925. The distribution of the alfalfa weevil (*Phytonomus posticus* Gyll.). A study in physical ecology. J. Agric. Res. 30: 479-491.

_____. 1929. A bioclimatic zonation for studying the economic distribution of injurious insects. Ecology 10: 282-293.

Dallwitz, M. J. & J. P. Higgins. 1978. Users guide to DEVAR: a computer program for estimating development rate as a function of temperature. Commonwealth Scientific and Industrial Research Organization, Division Entomol. Report No. 2.

Dennill, G. B. & A. J. Gordon. 1990. Climate-related differences in the efficacy of the Australian gall wasp (Hymenoptera: Pteromalide) released for control of *Acacia longifolia* in South Africa. Environmental Entomol. 19: 130-136.

Drake, A., H. A. Mooney, F. Di Castri, R. H. Goves, F. J. Kruger, M. Rejmanek & M. Williamson. 1989. *Biological Invasions. A Global Perspective*. Published for the Scientific Committee on Problems of the Environment, International Council of Scientific Unions. Wiley, New York.

Forer, P. & L. Chalmers. 1987. "Geography and Information Technology: Issues and Impacts," in P. G. Holland & W. B. Johnston, eds., *Southern Approaches: Geography in New Zealand*. Pp. 35-57. New Zealand Geographical Soc., Christchurch, New Zealand.

Frampton. E. R. 1990. "The Impact of Climate Change on Quarantine Pests," in R. A. Prestidge & R. P. Pottinger, eds., *The Impact of Climate Change on Pests, Diseases, Weeds and Beneficial Organisms Present in New Zealand Agricultural and Horticultural Systems*. Pp. 171-179. Report for New Zealand Ministry for the Environment. Plant Protection Group. Ruakura Agricultural Centre. Hamilton, New Zealand.

Gilpin, M. 1990. Ecological prediction. Science 248: 88-89.

Higley, L. G., L. P. Pedigo & K. R. Ostlie. 1986. DEGDAY: A program for calculating degree-days, and assumptions behind the degree-day approach. Environmental Entomol. 15: 999-1016.

Howe, R. W. 1957. A laboratory study of the cigarette beetle, *Lasioderma serricorne* (F.) (Coleoptera: Anobiidae) with a critical review of the literature on its biology. Bul. Entomological Research 48: 9-56.

Howe, R. W. & H. D. Burges. 1953. Studies on beetles of the family Ptinidae. 9. A laboratory study of the biology of *Ptinus tectus* Boield. Bul. Entomol. Research 44: 461-516.

Jordan, G. & L. Vietinghoff. 1987. "Fighting Budworm with a GIS," in *Proceedings, 8th International Symposium on Automated Cartography*. Pp. 492-499. American Society for Photogrammetry and Remote Sensing. Falls Church, Virginia.

Kaufmann, O. 1932. Einige Bemerkuungen über den Einfluss von Temperaturschwankungen auf die Entwicklungsdauer und Streuung bei Insekten und seine graphische Darstellung durch Kettelinie und Hyperbel. Zeitschrift für Morphologie und Ökologie der Tiere 25: 353-361.

Köppen, W. 1931. Grundriss der klimakunde. W. de Gruyter, Berlin.

Maywald, G. F. & R. W. Sutherst. 1989. User's guide to CLIMEX, a computer program for comparing climates in ecology. Commonwealth Scientific and Industrial Research Organization, Division Entomol. Report 2nd edition. 35.

McClain, D. C., G. C. Rock & R. E. Stinner. 1990. San Jose scale (Homoptera: Diaspididae): simulation of seasonal phenology in North Carolina orchards. Environmental Entomol. 19: 916-925.

Meats, A. 1981. The bioclimatic potential of the Queensland fruit fly, *Dacus tryoni*, in Australia. Proceedings, Ecological Soc. Australia 11: 151-161.

Messenger, P. S. 1964a. Use of life tables in a bioclimatic study of an experimental aphid-braconid wasp host-parasite system. Ecology 45: 119-131.

_____. 1964b. The influence of rhythmically fluctuating temperatures on the development and reproduction of the spotted alfalfa aphid, *Therioaphis maculata*. J. Econ. Entomol. 57: 71-76.

_____. 1970. "Bioclimatic Inputs to Biological Control and Pest Management Programmes," in R. L. Rabb & F. E. Guthrie, eds., *Concepts of Pest Management*. Raleigh, N.C.: North Carolina State University.

_____. 1974. "Bioclimatology and Prediction of Population Trends," in *Proceedings of the FAO Conference on Ecology in Relation to Plant Pest Control, Rome 1972.* Pp. 21-45. Food and Agriculture Organization of the United Nations, Rome.

_____. 1976. "Experimental Approach to Insect-Climate Relationships," in *Climate and Rice*. International Rice Research Institute, Los Banos, Philippines.

Messenger, P. S. & N. E. Flitters. 1954. Bioclimatic studies of three species of fruit flies in Hawaii. J. Econ. Entomol. 47: 756-765.

Mochida, O., R. C. Joshi & J. A. Litsinger. 1987. "Climatic Factors Affecting the Occurrence of Insect Pests," in *Weather and Rice*. Pp. 149-164. International Rice Research Institute, Los Banos, Philippines.

Montgomery, B. A. 1987. Gypsy moth in Michigan, the first annual report of the gypsy moth technical committee. Michigan Department of Agr. Lansing, Michigan.

North American Plant Protection Organisation. 1991. A process for analyzing the risk to domestic plants posed by a foreign biotic agent. A discussion. NAPPO Pest Risk Analysis Panel. August 1991.

O'Neill, R. V., R. H. Gardner & J. B. Mankin. 1980. Analysis of parameter error in a nonlinear model. Ecological Modelling 8: 297-311.

Palca, J. 1990. Libya gets an unwelcome visitor from the West. Science 249: 117-118.

Pielou, E. C. 1981. The usefulness of ecological models: a stocktaking.

Quarterly Review of Biology 56: 17-31.

Pimm, S. L. & D. P. Bartell. 1980. Statistical model for predicting range expansion of the red imported fire ant, *Solenopsis invicta*, in Texas. Environmental Entomol. 9: 653-658.

Podolsky, A. S. 1984. *New Phenology. Elements of Mathematical Forecasting in Ecology*. New York: John Wiley & Sons.

Pradhan, S. 1945. Insect population studies. II. Rate of insect development under variable temperature of the field. Proceedings, National Institute of Sciences of India. Part B Biological Sciences 11: 74-80.

Ratte, H. T. 1985. "Temperature and Insect Development," in K. H. Hoffman, ed., *Environmental Physiology and Biochemistry of Insects*. Pp. 33-66. New York: Springer-Verlag.

Ring, D. R., M. K. Harris, J. A. Jackman & J. L. Henson. 1983. A FORTRAN computer program for determining start date and base temperature for degree day models. Texas Agric. Experiment Station MP-1537, Texas.

Seamens, H. L. 1945. A preliminary report on the climatology of the wheat stem sawfly (*Cephus cinctus* Nort.) on the Canadian prairies. Scientific Agr. 25: 432-437.

Siddiqui, W. H. & C. A. Barlow. 1972. Population growth of *Drosophila melanogaster* (Diptera: Drosophilidae) at constant and alternating temperatures. Annals Entomological Soc. America 65: 993-1001.

Stiling, P. 1990. Calculating the establishment risks of parasitoids in classical biological control. American Entomologist 36: 225-230.

Stinner, R. E., G. C. Rock & J. E. Bacheler. 1988. Tufted apple budmoth (Lepidoptera: Tortricidae): simulation of postdiapause development and prediction of spring adult emergence in North Carolina. Environmental Entomol. 17: 271-274.

Stone, W. E. 1939. An instrument for the reproduction, regulation, and control of variable temperature. J. Washington Academy Sci. 29: 410-415.

Sutherst, R. W. & G. F. Maywald. 1985. A computerized system for matching climates in ecology. Agric. Ecosystems Environment 13: 281-299.

Tanigoshi, L. K., R. W. Browne, S. C. Hoyt & R. F. Lagier. 1976. Empirical analysis of variable temperature regimes on life stage development and population growth of *Tetranychus mcdanieli* (Acarina: Tetranychidae). Annals, Entomological Soc. America 69: 712-716.

Wagner, T. L., H-I. Wu, P.J.H. Sharpe, R. M. Schoolfield & R. N. Coulson. 1984. Modeling insect development rates: a literature review and application of a biophysical model. Annals, Entomological Soc. America 77: 208-225.

Worner, S. P. 1988a. Evaluation of diurnal temperature models and thermal summation in New Zealand. J. Econ. Entomol. 81: 9-13.

_____. 1988b. Ecoclimatic assessment of potential establishment of exotic pests. J. Econ. Entomol. 81: 973-983.

_____. 1991. Use of models in applied entomology: the need for perspective. Environmental Entomol. 20: 768-773.

_____. 1992. Performance of phenological models under variable temperature regimes: consequences of the Kaufmann or rate summation effect. Environmental Entomol. 21: 689-699.

3

Statistical Methods for Quarantine Treatment Data Analysis

Victor Chew

A common problem in statistics is to infer from a known sample some of the characteristics (e.g., means, variances, regression coefficients) of the unknown population from which the sample values have been drawn. Two kinds of statistical inference problems exist: hypothesis testing and interval estimation.

The conclusion from standard (nonsequential) hypothesis testing is to accept or reject H_o, the null hypothesis being tested (e.g., equality of two means or of two regression coefficients). A Type I error occurs if H_o is rejected when H_o is in fact true (false positive). A Type II error is committed if H_o is wrongly accepted (false negative).

Hypothesis testing is an academic exercise in futility since no two population means or regression coefficients are ever equal (Chew 1977). It is also of no practical consequence whether they are equal or not in the thousandth decimal place. A more informative alternative to testing is estimation of the difference by means of an interval.

Sample Size

Naturally, it is desirable to make the probability α of a Type I error and the probability β of a Type II error small. However, for a given sample size, α can be decreased only at the expense of increasing β, and vice versa. To decrease both α and β simultaneously, the sample size must be increased. There is a relationship between α, β, sample size, and standardized difference (expressed in units of the standard deviation) that can be used to calculate the number of replications such that the experiment will have a prescribed probability of detecting a given difference of practical importance (Chew 1984).

A different kind of sample size problem exists in quarantine studies. For a quarantine treatment to be acceptable, it must attain probit 9 efficacy, a survival rate of no more than 32 treated stages per million (Chew & Ouye 1985). The problem is the estimation of the true survival proportion of a treatment. Clearly, a point or single value estimate is not satisfactory. A treatment is 100% effective if, for example, one larva is treated and it does not survive. An interval estimate is needed here. Also, a one-sided upper bound of the form $(0,u)$ for the estimate of the survival proportion of the treatment is more informative than the two-sided confidence limits (L,U).

Couey & Chew (1986) gave formulas connecting the upper bound u, sample size N, number of survivors S, and the confidence probability γ. For example, if there is no survivor out of N = 35,000 larvae tested, the probability is γ= 0.95 that the unknown survival proportion is between: (a) L = 7 and U = 159 per million and (b) between zero and u = 136 per million. To be able to say, with the same confidence, that the true survival proportion is <136 per million is more informative for quarantine purposes than to say that it is between 7 and 159 per million. More precision is gained on the upper limit by sacrificing information on the lower limit, which is of no interest here.

Caution should be exercised in using the sample size formulas. In practice, there may be two serious violations of assumptions. Because of practical considerations, the required large number of larvae will have to be treated in several experiments, so that there will be an additional experiment-to-experiment variation, in addition to the variation among larvae in the same experiment. Secondly, the number treated is usually unknown and is estimated from an independent sample of artificially infested fruits where the larvae are allowed to emerge. It is impossible to derive sample size formulas to take into account these two additional sources of variation.

Experimental Design

The classical and popular factorial experiments have the disadvantage that the total number of runs increases rapidly as the number of factors increases. Furthermore, if the factors are quantitative (e.g., dose, time), the quadratic coefficients of the regression model are estimated poorly from a factorial experiment.

There is a class of designs called Rotatable Central Composite Designs (RCCDs) which could aid in performing experiments with >1 factor, including designs that permit blocking. RCCDs with up to six factors are tabulated in Cochran & Cox (1957). For example, a three-factor design consisting of 20 runs can be divided into three blocks of six, six, and eight runs, if one cannot finish all 20 runs in one day, or if there is not enough material from one batch for 20 runs. As with ordinary Randomized Complete Block Designs, block differences do not affect treatment comparisons. See Khuri & Cornell (1987) for an extensive discussion of other classes of designs for fitting response surfaces.

To avoid bias, it is important to do the experimental runs in a completely random order, if there is no blocking; with blocking, the runs within each block must be randomized. This may be very inconvenient. Thus, in an experiment to study the effects of immersion in hot water at different temperatures for different times, it will be very tempting to start with the lowest temperature and to remove the fruits in the order of increasing times, then proceed to the next higher temperature. Not only will bias be introduced through lack of randomization, but correlation will also be induced among data from the same temperature, through unwittingly performing a split plot experiment, with temperature as the main plot and time as the subplot treatments.

Simple Linear Regression

Simple linear regression analysis is the study of the relationship between a dependent variable y and an independent variable x. This can be extended to simple curvilinear regression, multiple linear regression (linear relationship between y and two or more independent variables), and multiple curvilinear regression (see appendix.)

The linear (straight line) regression model can be written as

$$y_i = a + bx_i + e_i \tag{1}$$

where a is the intercept (height of the line above the origin), b is the slope (increase in y per unit increase in x), and e_i is the error (sampling, measurement) in y_i, the i-th observation corresponding to x_i. In a typical application, y_i could be the proportion of larvae killed when exposed to dose x_i. The error term allows different batches of larvae to have different mortality rates when exposed to the same dose. The errors are assumed to be normally and independently distributed with a common variance. If the assumption of a constant variance is untenable, weighted regression analysis must be performed. With known weights, this is easy; quite often, however, the weights have to be estimated from the data and the computation then has to be iterated (see appendix.)

Linear relationships are the easiest to handle and interpret. They are all of the form given in (1), whereas there are infinitely many nonlinear relationships. If the relationship between y and x (in their original units) is nonlinear, a transformation of y and/or x will sometimes produce a linear relationship. For example, the equation $y = ax^k$ will be linear if we plot $\log(y)$ against $\log(x)$. $\log(y)$ versus x is linear if the relationship is of the form $y = ae^{-bx}$. The equation $y = ae^{-bx} + ce^{-dx}$ cannot be linearized.

The equation $y = a + bx$ and the covariance matrix of the coefficients summarize all of the information contained in the data. The equation can be used directly to predict y (e.g., mortality) at a given x (e.g., dose) and the confidence limits for the true y; or, inversely, to find x that will produce a given y (e.g., the dose that will kill 99% of the treated stage). Because b is the increase in y for every unit increase in x, it does not make sense to test if 20 and 30 parts per million (ppm), say, are different. In fact, 20 and 20.1 ppm will be different, by the amount of b/10. This difference may be negligible in practice, but the statistical test of significance tests if the true difference is zero, not if the difference is significant in the practical or economic sense (Chew 1980).

In comparing two lines $(a_1 + b_1x)$ and $(a_2 + b_2x)$, three cases may arise: (i) $a_1 = a_2$ and $b_1 = b_2$ (the two lines are coincident and the two treatments are equal at all doses); (ii) $b_1 = b_2$ but a_1 is not equal to a_2 (the two lines are parallel and one treatment is better than the other at all doses); or (iii) b_1 and b_2 are unequal. In this last case, the two lines will intersect at an unknown dose x^*. If x^* is within the experimental range, then one treatment will be superior at doses below x^* and the other will be superior at doses above x^*; if x^* is beyond the experimental range of doses, then within this range, one treatment is superior to the other. In none of the three cases above does it make sense to ask if two treatments are equal at a particular dose.

Dose-Mortality Relationship

A very common problem in quarantine research is the estimation of dose-mortality relationships. A known number of subjects (larvae, fruits) are exposed to each of several levels of stimulus (dose, temperature, time) and the number responding (death, decay) at each dose is recorded. It is assumed that each subject has a tolerance level such that it will die if exposed to a dose exceeding its tolerance. Usually, this tolerance cannot be measured; otherwise, from a sample of subjects, the mean and variance of the distribution of tolerances can be estimated. If, additionally, this distribution is assumed to be normal, it is possible to predict the percentage mortality at any given dose, or to find that dose which will kill 99.9968%, (probit 9), of the population of subjects.

If, in the regression equation (1), y is the proportion mortality and x is dose, the plot of y versus x will not be linear. If the distribution of tolerances is normal, it can be shown that the probit transformation of y will be linearly related to x (Finney 1971). There is no formula that can be used to calculate the probit transformation of a given proportion. It is only defined implicitly as a definite integral of the standard normal distribution (zero mean and unit standard deviation). The probit transformation of a proportion p is that z value of the standard normal distribution such that the area to the left of that z value is p. The following short excerpt from the z table will illustrate the probit transformation.

proportion (p):	0.01	0.05	0.50	0.975	0.99	0.999968	
probit of p (z):	−2.33	−1.64	0.00	1.96	2.33	4.000000	(2)

The probit transformation of a proportion that is <0.50 is negative. To avoid these negative probits, five is added in the definition (Finney 1971). (A proportion will be virtually zero for its probit transformation to be <−5.0.) Since the probit transformations of zero and one are undefined, McCullagh & Nelder (1989) modified the proportion killed as $(r + 0.5)/(n +1)$, if r died out of n subjects tested.

If the tolerance distribution is lognormal instead of normal, the doses will have to be transformed into logarithms in order to achieve linearity. Since it is not possible to determine whether the tolerance distribution is normal or lognormal, the obvious suggestion is to try both and see which assumption gives a better fit to the data, as measured by the lack-of-fit chi-square test.

In addition to the normal, two other distributions (logistic and Gompertz) are sometimes assumed for the tolerance distribution, leading respectively to the logit and ln(−ln) linearizing transforms of the proportion mortality p, where ln is the natural logarithm.

logit(p)	$= \ln \{p/(1-p)\}$	(3)
Gompertz(p)	$= \ln \{-\ln(1-p)\}$	(4)

Only a pocket calculator is needed to get the logit and Gompertz transformations of p. (The Gompertz transformation is sometimes called the complementary log-log transformation. Altenburg & Rosenkranz

(1989) called it the Weibull transformation, while Collett (1991) referred to it as the Gumbel transformation.)

There is a problem in fitting the regression line. The variance of y_i (probit, logit, or Gompertz transformation of proportion mortality) is unknown as well as not being constant. It involves the unknown true proportion mortality. The computations have to be iterated. In the first iteration, the unknown true proportion mortality is estimated by the observed proportion mortality, and a weighted regression analysis is performed. In the second iteration, a better estimate of the unknown true proportion mortality is obtained from the regression line, and a second weighted regression line is fitted. The process is repeated until there is no appreciable change in the error mean square or in the estimates of the slope and intercept.

In an analysis of variance of proportions, a common transformation is the angle (arcsine of the square root of p). An advantage of this transformation is that the variance of the angle (in radians) is equal to $1/(4n)$, which is constant if n is constant for all treatments (requiring no weighted analysis) or, at worst, it is known, so that no iteration is necessary. This transformation has been used for bioassays (Knudsen & Curtis 1947, Claringbold et al. 1953). "For all practical purposes, the angle transformation is, therefore, a linear function of the Probit transformation" (Shuster & Dietrich 1976).

There is no theoretical basis for choosing among the four transformations. These have been compared by Naylor (1964). Cox & Snell (1989) found that the logistic and the normal (probit) agree closely over the whole range except for the region where the probability of success is very near one. Then the normal curve approaches its limit more rapidly than the logistic. A different conclusion was obtained by Agresti (1990) who analyzed a set of published data using the probit, logit, and Gompertz transformations, and found that the Gompertz model fits better than logit and probit models. Besides being computationally more convenient than probits, logits have the advantage of being more interpretable. The quantity $p/(1-p)$ is the ratio of the probability of success to that of failure (called the odds ratio) so that the regression equation, in terms of logits, shows how the logarithm of the odds ratio changes with the independent variable x. Unlike probits and logits, the complementary log-log transform is not symmetrical about $p = 0.5$. Collett (1991) discussed situations where this transformation arises naturally (dilution assays, grouped survival data analysis, serological testing, and reliability analysis).

Once the regression equation has been obtained, it can be used to find the probit (or logit) of the proportion that will be killed by a given dose x* (probit or logit = a + bx*), and from z tables or equation (3), find the proportion corresponding to this probit or logit. Confidence limits on the unknown true proportion can be calculated in the usual way. The regression equation can also be used inversely; for example, to find LD_{99} with corresponding probit of 2.33, solve 2.33 = a + bx, giving x = (2.33 − a)/b. Confidence limits for the true LD_{99} are more difficult to obtain. SAS gives the values of LD (100p) and their limits for p = 0.01 − 0.99, in steps of 0.05 between 0.10 and 0.90 and in steps of 0.01 elsewhere. A SAS/PC program for calculating limits for other values of p is available from M. J. Firko, USDA-APHIS-BBEP, 6505 Belcrest Road, Hyattsville, MD 20792.

The relative potency of two treatments is defined as the ratio of equally effective doses and can be obtained analytically or graphically (Finney 1971). For the latter a horizontal line is drawn. Where this line intersects the regression lines of the two treatments, perpendicular lines are drawn down to the equally effective doses. Limits for this relative potency can be calculated. Because the regression lines are assumed to be parallel, it does not matter whether the horizontal line goes through, for example, 50 or 80% mortality; otherwise, there will be a different relative potency for each percent mortality.

Questions about sample size and experimental design are very hard to answer, unlike the case where only one parameter (a mean or a proportion) is being estimated. In a dose-mortality study, there are at least four unknowns of interest: intercept, slope, proportion p killed at a given dose x, and dose x that will kill a given proportion p. The last two contain an infinite number of unknowns, because of the infinite number of given doses and proportions. A design that is optimal for estimating LD_{50} is not optimal for estimating LD_{99}. The variance of the estimated unknown is a function of the design points (including sample size) multiplied by the error variance. The design points are independent of data, and the error variance is uncontrollable. Therefore, at the design stage, two competing designs can be compared on the basis of the variances of the estimates from these two designs. Once the design proper has been chosen (number of doses and values of the doses), the sample size question can be answered if the required precision of the estimate is specified.

There has been much discussion on the appropriateness of probit 9 as a quarantine requirement (Landolt et al. 1984). It does seem strange to require 99.9968% mortality, instead of rounding it to 99.0, 99.9, or 99.99. The actual protection level depends on several factors. If fruits are highly infested, a treatment with a very high mortality rate is clearly needed; on the other hand, if fruits come from a pest free zone, the treatment efficacy can be relaxed and still maintain quarantine security.

For example, if 1,000,000 fruits (with an infestation rate of one larva per 1,000 fruits) are treated with a treatment that has 99.90% mortality rate, the expected number of larvae surviving the treatment is 1,000,000(0.001)(0.001) = 1.

Computer Software

All of the major statistical analysis software packages have the capability of analyzing quantal (success/failure) data: SAS (SAS Institute Inc., Cary, NC 27512), BMDP (Biomedical Package, BMDP Statistical Software Inc., Sepulveda Blvd., Los Angeles, CA 90025), SPSS (Statistical Package for Social Scientists, SPSS Inc., N. Michigan Ave., Chicago, IL 60611), EGRET (Epidemiological, Graphics, Estimation and Testing, Statistics and Epidemiology Research Corp., 909 NE 43rd St, Seattle, WA 98105), GLIM (Generalized Linear Interactive Modelling), and Genstat (General Statistical) program. The last two are British packages, obtainable from NAG Inc., 1400 Opus Place, Suite 200, Downers Grove, Chicago, IL 60414. Besides these packages (also available for PCs), there are also special programs: POLO2 (Robertson et al. 1981), PRODOS (Ihm et al. 1987), PCPROBIT (Walsh 1987).

The probit procedure in SAS/PC allows the three transformations (probit, logit, and Gompertz), with or without the log transformation of the dose (SAS Institute 1988, 1990). It also has the OPTC option which optimizes C, natural mortality, for a total of 12 possible analyses. If OPTC is not specified, SAS takes C = 0 or keeps C unrevised in the iterations, depending on whether control data are absent or present, respectively. If OPTC is specified, SAS revises the estimate of C in each iteration, and reduces the error degrees of freedom by one. SAS does not estimate the natural immunity, the proportion of subjects that will not respond to the stimulus, no matter how high the level of that stimulus. It ignores what Finney (1971) called Wadley's problem: the number of larvae in the fruits exposed at each dose is often unknown and estimated from a parallel sample of untreated fruits.

Besides estimates of slope and intercept (and their covariance matrix), SAS also prints out the mean μ and standard deviation (σ) of the tolerance distribution and their covariance matrix. These two sets are mathematically related:

intercept = $-\mu/\sigma$ (add five to numerator if this is added in the definition of probit);
slope = $1/\sigma$, so that regression line will be steep if the σ of the tolerance distribution is small (fairly homogeneous subjects).

SAS does not calculate relative potency. Altenburg & Rosenkranz (1989) gave the macro listing for doing this in SAS. SAS allows two or more independent variables. It can also handle more than two categories of responses (e.g., alive, moribund, and dead). Some information is lost in grouping the responses into two classes, as in grouping alive and moribund into one class.

In cases of a poor fit, SAS (following Finney 1971) multiplies the variances and covariance of the estimates by a heterogeneity factor (the goodness-of-fit chi-square statistic divided by its degrees of freedom), and uses the Student's t-distribution instead of the standard normal distribution in calculating the fiducial limits. This method is only valid when the binomial denominators are all equal. Circumstances in which each proportion is based on the same number of binary observations are comparatively rare. However, the procedure is not too sensitive to differences in the values of n_i, and so this method of allowing for over-dispersion can be used as a first approximation even when the n_i are not all equal (Collett 1991). This reference also discussed four other alternative methods for handling lack of fit. SAS cannot perform these alternative methods. GLIM and EGRET can do some of them.

Besides the probit, logit, and Gompertz models, an alternative is to try other tolerance distributions. These have been suggested by Prentice (1976), Copenhaver & Mielke (1977), Guerrero & Johnson (1982), Morgan (1985), and Stukel (1988). Aranda-Ordaz (1981) proposed the following transformation of p:

$$g(p) = \ln[\{(1-p)^{-\lambda} -1\}/\lambda] \qquad (5)$$

which becomes the logistic if $\lambda = 1$ and tends to the Gompertz as λ tends to zero. Other values of λ may fit the data if the logistic and Gompertz do not. A program for this transformation called QUAD (Quantal Assay Data) is being developed by B.J.T. Morgan of the Mathematical Institute of the University of Kent at Canterbury, England. In the meantime, trial values of λ may be used; e.g., $\lambda = 0.10 - 0.90$. Without the logarithm, equation (5) is the Box-Cox transformation (Draper & Smith 1981) that is often used in analyses of variance for variance stabilization.

Appendix: Iteratively Reweighted Least Squares

The most general linear regression model is

$$y_i = b_0(x_{0i}) + b_1(x_{1i}) + b_2(x_{2i}) + \ldots + b_p(x_{pi}) + e_i \qquad (6)$$

where y_i is the i-th observation taken at $x_1 = x_{1i}, \ldots,$ and $x_p = x_{pi}$; x_0 is a dummy variable and $x_{0i} = 1$ for all values of $i = 1, \ldots, n$. It includes as particular cases simple linear regression ($p = 1$); simple curvilinear regression (with $x_k = x^k$; multiple linear regression ($x_1, \ldots, x_p$ represent p different experimental factors); and multiple curvilinear regression with k factors, where k is less than p, and there are squared and/or cross product terms in (6). For example, with x_1 and x_2 denoting the two factors, $x_3 = x_1{}^*x_2$, $x_4 = x_1{}^*x_1$, and $x_5 = x_2{}^*x_2$, (6) is the second degree response surface model in two independent variables. In (6), x is a known constant (the experimental settings of the p independent variables). Regression coefficients (b) are the unknowns. Equation (6) is linear in the unknowns and is called a linear model, even if the relationship between y and the x values is nonlinear, as in two out of the four cases above. For an example of nonlinear and nonlinearizable models, see the section on simple linear regression.

The output from a regression analysis may include several different types of sums of squares. The most important are Type I (or sequential) SS and Type III (or partial) SS. These SSs, in general, are different as they test different hypotheses, except for simple linear regression. Use Type I for simple curvilinear regression and Type III for multiple linear regression. For multiple curvilinear regression, look at both types of SSs, keeping in mind that Type I SS for any regressor variable is adjusted or corrected for only those variables that precede it in the model, while the Type III SS for that variable is adjusted for all other variables in the model. The Type I SS for a variable, therefore, depends on its position in the model, while the Type III SS is independent of the ordering of the variables in the model.

Equation (6) can be expressed very concisely in matrix and vector notation as

$$y = X b + e \qquad (7)$$

where y is an (n x 1) column vector of observations, b is a (p+1) x 1 column vector of regression coefficients, e is an (n x 1) vector of errors in y, and X is an n x (p+1) matrix of known constants. The matrix X is appropriately called the design matrix, since the rows of X give the settings of the p factors in each of the n runs. Equation (6) is the i-th row of (7). Without the dummy variable x_0, (6) cannot be written as concisely as in (7). Matrix algebra is almost essential in discussing regression analyses. Many statistics textbooks have a chapter or an appendix on matrix algebra (Draper & Smith 1981).

If S is the covariance matrix of e (i.e., an (n x n) matrix whose i-th diagonal element is the variance of e_i and whose (i,j)-th element gives the covariance of e_i and e_j, assumed known completely), then from Gauss-Markov Theorem, the best (minimum variance unbiased) estimate of b in (7) is

$$b = (X^T S^{-1} X)^{-1} (X^T S^{-1} y) \tag{8}$$

where the superscripts T and -1 denote transpose and inverse, respectively. The same letter b has been used for convenience to denote the vector of unknown regression coefficients (7) and its estimate (8). It should be clear from the context which is meant. Note that b in (8) is unchanged if each element in S is multiplied or divided by some constant c. The covariance matrix of b is

$$\text{cov.}(b) = (X^T S^{-1} X)^{-1} \tag{9}$$

a $(p+1) \times (p+1)$ matrix. The elements of cov.(b) are used to test hypotheses about the regression coefficients and to construct confidence limits for regression coefficients, predicted values, etc. These tests and limits will be based on the standard normal or z-distribution, since the variances are assumed known. The above case of known S is virtually nonexistent. At best, we know S up to an unknown constant multiplier; i.e.,

$$S/\sigma^2 = V \tag{10}$$

where V is completely known but σ^2 is unknown. The estimate of b is still given by (8), with unknown S replaced by known V; however, the covariance matrix is now

$$\text{cov.}(b) = (X^T V^{-1} X)^{-1} \sigma^2 \tag{11}$$

where σ^2 is estimated by the error mean square in the regression analysis. Tests and confidence limits will now be based on the Student's t-distribution, with degrees of freedom equal to that of the error mean square. This case of known $V = S/\sigma^2$ but unknown S and σ^2 occurs when n_i replicates are run in the i-th experiment, and the averages are analyzed. Here y_i is the average of n_i observations, and the variance of its error e_i is σ^2/n_i. The covariance matrix S of the error vector is of the form $S = V\sigma^2$.

Confidence limits for the regression coefficients and predicted values are based on (11). The factor σ^2 (error mean square) is independent of the design. The other factor depends only on the design and not on the data. Thus at the design stage if there are two competing designs, (11) can be calculated, without σ^2, to determine which design gives the smaller variance. Unfortunately, the design that is optimum for estimating one parameter may not be optimum for estimating some other parameter.

The final case to be considered is where the errors are uncorrelated but the variances are unequal and unknown, as in probit, logit, or Gompertz analysis. The covariance matrix S is diagonal. If W is the inverse of the diagonal matrix S, W will also be a diagonal matrix, and its diagonal element w_i (called the weight of y_i) is the reciprocal of v_i (variance of y_i). Also, only simple linear regression will be considered. The formulas for the slope b and intercept a are

$$b = [(\Sigma w)(\Sigma wxy) - (\Sigma wx)(\Sigma wy)]/[(\Sigma w)(\Sigma wx^2) - (\Sigma wx)^2] \qquad (12)$$

$$a = (\Sigma wy)/(\Sigma w) - b(\Sigma wx)/(\Sigma w) \qquad (13)$$

where Σwxy, for example, is the summation of wxy; i.e., $\Sigma wxy = w_1 x_1 y_1 + \ldots + w_n x_n y_n$. If the variance of y_i is constant, its weight may be taken as unity, and the above formulas reduce to the usual formulas for slope and intercept.

No iterative computations are needed in doing a weighted regression analysis if the weights w_i are known. If p is the observed proportion mortality based on n subjects, then from the binomial distribution, the variance of p is $P(1-P)/n$, where P is the true (unknown) mortality proportion. The variances of the probit, logit, and Gompertz transformations of p can be obtained from the covariance propagation theorem (sometimes called the delta method) and are given below. (The variance of the arcsine transform has been given previously.)

$$\text{var.(probit of } p) = P(1-P)/(nh^2) \qquad (14)$$

$$\text{var.(logit of } p) = 1/\{nP(1-P)\} \qquad (15)$$

$$\text{var.(Gompertz of } p) = P/[n(1-P)\{Ln(1-P)\}^2] \qquad (16)$$

In (14), h is the ordinate of the standard normal curve corresponding to P.

In the first iteration, P is approximated by p. The variances and hence the weights are now known. Applying formulas (12) and (13), fit a weighted regression line. Specifically, if the logit transformation is used, so that logit = a + bx, the logit at each of the doses is calculated.

Back transformations are made to convert these logits into proportions. From Equation (3), $p = 1/(1 + exp(-logit of p))$. These proportions (calculated from the regression line) are now taken as P in the second iteration. The iterations are repeated until a given convergence criterion is met. Because the weights change at each iteration, the procedure is called iteratively reweighted least squares.

References

Agresti, A. 1990. *Categorical Data Analysis*. New York: Wiley.

Altenburg, H.-P. & G. Rosenkranz. 1989. Analysis of dose response data with SAS. Fakultat fur Klinische Medizin Mannheim der Universitat Heidelberg, Med. Statistik, Biomathematik und Informations-verarbeitung, Mannheim, Germany and Hoechst AG, Pharma Forschung Informatik, D-6230 Frankfurt/M 80, Germany.

Aranda-Ordaz, F. J. 1981. On two families of transformations to additivity for binary response data. Biometrika 68: 357-363.

Chew, V. 1977. Statistical hypothesis testing: an academic exercise in futility. Proceedings, Fla. State Hortic. Soc. 90: 214-215.

_____. 1980. Testing differences among means: correct interpretation and some alternatives. HortScience 15: 467-470.

_____. 1984. Number of replicates in experimental research. The Southwestern Entomologist Supplement No. 6. Pp. 2-9.

Chew, V. & M. T. Ouye. 1985. "Statistical Basis for Quarantine Treatment Schedule and Security," in J. H. Moy, ed., *Radiation Disinfestation of Food and Agricultural Products, Proceedings of an International Conference, Honolulu (1983)*. Pp. 70-74. Honolulu, Hawaii: Hawaii Institute of Tropical Agr. and Human Resources, University of Hawaii at Manoa.

Claringbold, P. J., J. D. Biggers & C. W. Emmens. 1953. The angular transformation in quantal analysis. Biometrics 9: 467-484.

Cochran, W. G. & G. M. Cox. 1957. *Experimental Designs, 2nd Ed.* New York: Wiley.

Collett, D. 1991. *Modelling Binary Data*. New York: Chapman and Hall.

Copenhaver, T. W. & P. W. Mielke. 1977. Quantit analysis: a quantal assay refinement. Biometrics 33: 175-186.

Couey, H. M. & V. Chew. 1986. Confidence limits and sample size in quarantine research. J. Econ Entomol. 79: 887-890.

Cox, D. R. & E. J. Snell. 1989. *Analysis of Binary Data, 2nd Ed.*, New York: Chapman and Hall.

Draper, N. & H. Smith. 1981. *Applied Regression Analysis, 2nd Ed.*, New York: Wiley.

Finney, D. J. 1971. *Probit Analysis, 3rd Ed.* Cambridge University Press.

Guerrero, M. & R. A. Johnson. 1982. Use of the Box-Cox transformation with binary response models. Biometrika 65: 309-314.

Ihm, P., H.-G. Müller & T. Schmitt. 1987. PRODOS: Probit analysis of several qualitative dose-response curves. The American Statistician 41: 79.

Khuri, A. I. & J. C. Cornell. 1987. *Response Surfaces: Designs and Analysis.* New York: Dekker.

Knudsen, L. F. & J. M. Curtis. 1947. The use of the angular transformation in biological assays. J. American Statistical Association 42: 889-902.

Landolt, P. J., D. L. Chambers & V. Chew. 1984. Alternative to the use of probit 9 mortality as a criterion for quarantine treatments of fruit fly (Diptera: Tephritidae)-infested fruit. J. Econ. Entomol. 77: 285-287.

McCullagh, P. & J. A. Nelder. 1989. *Generalized Linear Models, 2nd Ed.,* New York: Chapman and Hall.

Morgan, B.J.T. 1985. The cubic logistic model for quantal assay data. Applied Statistics 34: 105-113.

Naylor, A. F. 1964. Comparisons of regression constants fitted by maximum likelihood to four common transformations of binomial data. Annals of Human Genetics 27: 241-246.

Prentice, R. L. 1976. A generalization of the probit and logit methods for dose response curves. Biometrics 32: 761-768.

Robertson, J. L., R. M. Russell & N. E. Savin. 1981. POLO2: A computer program for multiple probit or logit analysis. Bul. Entomological Soc. America 27: 210-211.

SAS Institute, Inc. 1988. SAS Technical Report P-179, Additional SAS/STAT Procedures, Release 6.03. SAS Institute, Cary, North Carolina.

_____. 1990. SAS Technical Report P-200, SAS/STAT Software: CALIS and LOGISTIC Procedures, Release 6.04. SAS Institute, Cary, North Carolina.

Shuster, J. J. and F. H. Dietrich. 1976. Quantal response assays by inverse regression. Communications in Statistics, Part A - Theory and Methods 1: 293-305.

Stukel, T. A. 1988. Generalized logistic models. Jour. American Statistical Association 83: 426-431.

Walsh, D. 1987. PCPROBIT: A user-friendly probit analysis program for microcomputers. The American Statistician 41: 78.

4

Statistical Analyses to Estimate Efficacy of Disinfestation Treatments

*Jacqueline L. Robertson, Haiganoush K. Preisler,
E. Ruth Frampton, and John W. Armstrong*

In quarantine entomology, statistical analyses are used to estimate the probability that a commodity treatment will succeed. Laboratory bioassays are completed first, followed by a large-scale confirmatory test to validate the estimated treatment (e.g., dose, time) suggested as most suitable by the results of the laboratory trials. This sequence of testing is the same as that used in other entomological specialty areas concerned with the control of a pest species on a particular crop or in a particular environment. For example, large-scale trials of a chemical or microbial pesticide to control a pest species in a crop or in the forest environment are usually based on data provided from laboratory bioassays (Podgwaite et al. 1991, Weissling & Meinke 1991).

Compared with other research concerning arthropod control, however, the transition between laboratory bioassays and confirmatory tests is more direct in quarantine entomology. Operational conditions can be simulated reasonably well with small-scale fumigation chambers (Spitler & Couey 1983), hot forced air cabinets (Armstrong et al. 1989, Gaffney & Armstrong 1990, Sharp et al. 1991), vapor heat chambers (Gaffney et al. 1990) and other facilities. Thus, treatment rates selected in the laboratory can be translated almost directly to operational use without the need for use of conversion factors such as those derived by Haverty & Robertson (1982).

Here, we review the statistical methods used in commodity treatment bioassays that are done in the laboratory and that serve as the basis for confirmatory tests. In particular, we address problems of experimental design and data interpretation inherent in such bioassays. We summarize the statistical bases of confirmatory tests and the concept of the maximum pest limit. Finally, we discuss criteria that should be used by regulatory agencies to ensure the statistical validity of laboratory experiments used to establish requirements for disinfestation treatment of imported commodities.

Probit 9 Security:
A Scientific Anachronism

Successful treatment in quarantine entomology is one of the most rigidly defined of any specialized type of arthropod control. In 1939, Baker (1939) stated that "The security demanded as a basis for recommendation is determined by reading on the regression line the exposure coordinated with a probit of 9. In percentages, this probit represents a mortality of 99.99683%, or a survival of approximately 32 out of 1,000,000." The basis for Baker's recommendation seems quite simple: like many of his entomological contemporaries, he probably considered probit analysis as the standard method for the analysis of dose-response data. Based on knowledge available in 1939, he also concluded that 99.99683% mortality would ensure that the risk of accidental introduction of a pest would be virtually nonexistent. However, knowledge of exotic pests has increased substantially since 1939 (Armstrong & Couey 1989, Landolt et al. 1984) and a growing body of evidence suggests that mortality is far too narrow a criterion upon which to evaluate treatment efficacy (Baker et al. 1990, Landolt et al. 1984).

Baker's (1939) recommendation, like any other, should have been subjected to revision as knowledge about particular insect pests increased, and as entomologists became more familiar with biostatistics. Instead, probit 9 was codified as the requirement for quarantine security by the United States (U.S.) Department of Agriculture and comparable regulatory agencies of other nations. In the U.S. and most other countries that import commodities potentially contaminated with exotic pests, the requirement for probit 9 security has not changed despite 50 years of additional research in entomology and biostatistics.

Assumptions of the Probit 9 Requirement

Three interrelated assumptions are inherent in the probit 9 requirement. First, 99.9968% effectiveness is assumed to be the minimum level necessary for commodity protection. Second, the requirement implies that the probit model is always suitable for the analyses of data from commodity treatment bioassays. Finally, the probit 9 requirement assumes that no criterion other than death is relevant to the future establishment of the pest species in a new environment.

A more general definition of quarantine security is necessary to avoid the assumptions inherent in the probit 9 requirement as stated by Baker (1939). As described in the next section, laboratory bioassays for commodity treatments are merely quantal response bioassays, and all quantal response bioassays have the same statistical characteristics.

Quantal Response Bioassays in the Laboratory

A quantal response bioassay is any experiment in which the response of the test organisms varies in relation to a measurable characteristic of a stimulus (Robertson & Preisler 1992). In quarantine entomology, the stimulus may be a fumigant, exposure to low or high temperature, exposure to an atmosphere with altered gaseous composition, or other treatment. Experiments in quarantine entomology are done to quantify death of the test organism because mortality is the criterion specified by the probit 9 security requirement.

All quantal response bioassays differ in terms of the response variables (dependent variables) and the explanatory variables (independent variables) involved. Explanatory variables are the measurable characteristics of a stimulus or stimuli (e.g., concentration of a chemical, temperature) that cause responses of the target arthropod species to vary. Response variables are the random outcomes (e.g., death or survival) of the exposure to the stimulus: they vary in relation to the intensity of the stimulus.

Robertson & Preisler (1992) described various types of quantal response bioassays that can be done. The simplest type (and the type most often used in quarantine entomology) is a binary response experiment with one explanatory variable. In the rare instances in which more than one variable has been tested in a commodity treatment bioassay, the variables have been reduced into a single variable by multiplication of one variable by the other (e.g., the concentration times time [c x t] product [Monro 1969]).

As long as probit 9 security continues to be required by regulatory agencies, the advantages of performing bioassays with multiple explanatory variables will probably continue to be disregarded despite the fact that they can provide valuable information about variables that affect the probability that an exotic pest might become successfully established. Among the relevant variables are survival, level of infestation of the pest when the commodity is harvested, and suitability of a particular species or cultivar as a host for the quarantine pest (Armstrong 1985, Armstrong & Couey 1989).

Binary Quantal Responses

In the following sections, we review the factors that affect precise estimation of efficacy in binary quantal response bioassays, especially bioassays done to estimate a treatment level that will cause 99.9968% mortality. We describe the need for a more general terminology as a replacement for the probit 9 to describe the effects of quarantine treatments.

The statistical statement of binary response with a single explanatory variable is shown in Equation 1:

$$p_i = F(\alpha + \beta x_i) \tag{1}$$

p_i is the probability of response (e.g., death), x_i is the ith treatment level (e.g., dose) or a function of that level (e.g., logarithm of dose), α is the intercept of the regression line, β is the slope of the regression line, and F is a distribution function describing the shape of the response curve.

Transformation of the units of both the response variable and the explanatory variable are frequently (but not always) necessary for the relative frequency of responses to fit a linear model, despite Baker's (1939) statement that "The use of probits and logarithms or a logarithmic scale, as here used, converts the relationship between mortality and length of exposure to a linear one . . . ." Although units of the explanatory variable are usually converted to logarithmic values, this transformation is not always necessary for data to fit a line. For example, units of temperature or radiation tested in commodity treatments often need not be transformed at all for the data to fit the linear model (e.g., Burditt & Hungate 1988).

In binary quantal response bioassays, the points along the regression line are such that, for a given level of the explanatory variable, there is a corresponding probability of response. Examples of the level of the corresponding probability of the response variable are the LD (lethal dose), LC (lethal concentration), or LT (lethal time). For example, the LD_{50} is the dose estimated to be lethal to 50% of the population tested. In terms of lethal doses, the probit 9 criterion corresponds with the $LD_{99.9968}$.

The way a particular bioassay should be done (i.e., the experimental design) to estimate any lethal level depends on the response level of interest. The level necessary for 99.9968% mortality is especially difficult to estimate with any degree of certainty (regardless of the experimental design) because of its location in the upper extreme of the probability distribution (Copenhaver & Mielke 1977). In addition, the numerical value that results from the estimation procedure depends on the statistical model used: the probit model is only one of several available.

Probit or Logit Analysis

Of the numerous types available, the two distribution functions most commonly used in Equation 1 are the normal and the logistic curves (Robertson & Preisler 1992). The normal curve is assumed in probit analysis; the model specified by Equation 1 is $p_i = \Phi(\alpha+\beta x_i)$ where F is Φ, the standard normal (Gaussian) distribution function. For the logit model

$$p_i = \frac{1}{1+e^{-(\alpha+\beta x_i)}} = \frac{e^{\alpha+\beta x_i}}{1+e^{\alpha+\beta x_i}} \tag{2}$$

where e (= 2.71828) is the base of the natural logarithm.

Although a debate about use of one model versus the other has continued among statisticians (Berkson 1951, Finney 1964, Copenhaver & Mielke 1977), data from a binary quantal response bioassay are not always distributed as either a normal or a logistic function. Whether use of one model or the other is appropriate is difficult to determine because similar results are obtained with either one except at the extreme ends of the probability distribution. The difficulties involved in this problem provide an ample basis for our recommendation that more generalized terms be used in quarantine entomology.

In many instances, both the probit or logit model seems to fit bioassay data adequately. However, an adequate fit does not necessarily mean the best fit in the sense that another model that was not examined may be more appropriate for a particular data set. In any bioassay, results reflect both the statistical model and the assumptions inherent in the model.

Dose Selection and Sample Size

Two aspects of experimental design that are crucial in any binary response bioassay with one explanatory variable are dose selection and sample size. These two factors have primary effects on the precision of estimation. Before the study of Robertson et al. (1984), many researchers routinely used the guidelines for experimental design of dose-response bioassays described by Finney (1971) without realizing that these recommendations for dose selection pertain to estimation of the LD_{50} and not necessarily to other LDs. The applicability of the methods described by Brown (1966) and Freeman (1970) also are limited to the 50% response level or to responses in that vicinity. Tsutakawa (1980) presented another approach to the problem of dose selection for efficient estimation of an arbitrary lethal dose, but his method is not general and its computational difficulty precludes routine use. Thus, guidelines for the numbers of subjects to test at each dose and for the experiment as a whole were needed.

Empirical evidence (Haverty & Robertson 1982) suggested that different designs were required for precise estimation of the LD_{50} and the LD_{90}. A Monte Carlo computer simulation was used to test the effects of sample size and dose placement on the precision of LD_{50} and LD_{90} estimates and to find the optimal percentage of mortality that should be observed for three to eight doses in experiments with total sample sizes of 60 to 720 (Robertson et al. 1984). The criterion used for precision was the width of the 95% confidence limits around an LD_{50} or LD_{90} estimated with the logit model. This investigation suggested that at least 120 test subjects are required for a reliable dose-response experiment. Use of 60 to 64 test subjects (the smallest sample size that was tested) does not seem desirable because of the poor precision of the LD_{50} and LD_{90} estimates and the frequent occurrence of infinite 95% confidence limits (equivalent to no confidence limits at all). Because infinite 95% confidence limits occurred even with a sample size of 120, Robertson et al. (1984) concluded that at least 240 test subjects should be used to estimate a dose-response relationship with acceptable precision.

Although optimal sample size requirements for bioassays done to estimate an $LD_{99.9968}$ or $LT_{99.9968}$ have not been determined, we suspect that far more than 30,000 test subjects may be required based on the experience of the U. S. National Center for Toxicological Research in a study ("Megamouse") done to estimate the ED_{01} (effective dose to kill 1%) for a carcinogen, 2-acetylaminofluorene, to laboratory mice (Ottoboni 1984). In a megamouse study, a sample size of 24,000 was used and the result was still not definitive. Among other problems, no way exists to select the model that would be most appropriate for the data analysis. An analogous problem also occurs in quarantine entomology and is one which investigators must recognize from the outset of any experiment with a possible quarantine treatment.

Optimal placement of doses varies depending on the lethal dose of interest and the number of doses to be tested. For example, Robertson et al. (1984) found that precisely estimated LD_{50} values are obtained when responses are evenly distributed between 25 and 75%. Precise LD_{90}

estimates, however, require that one or two doses cause at most 10% mortality and that most doses cause between 75 and 95% mortality. The placement of treatment levels necessary to estimate $LD_{99.9968}$ or $LT_{99.9968}$ has not been determined because no study has shown that the probit model can always be assumed to be the best and because placement depends so heavily on the model assumed to be appropriate (Table 4.1).

TABLE 4.1. Comparison of estimates from probit and logit models with experimental data sets for *Bactrocera dorsalis* (C30, C31),[a] *Ceratitis capitata* (Mango),[b] and *Choristoneura occidentalis* (CO)[c]

Data	n[d]	Model[e]	Max[f]	LD or $LT_{99.9968}$ (95% CL)	LD or LD_{90} (95% CL)	X^2 (df)
C30	67,473	P	9	8.94 (8.12-9.76)	3.27 (3.04-3.51)	129 (7)
		L		11.31 (9.91-12.71)	3.07 (2.80-3.34)	180 (7)
C31	28,047	P	8	7.79 (5.62-9.95)	2.82 (2.33-3.30)	463 (6)
		L		8.85 (6.77-10.93)	2.81 (2.50-3.11)	229 (6)
Mango	24,451	P	70	76.1 (66.0-86.1)	37.1 (33.7-40.5)	246 (6)
		L		101.9 (85.5-118.3)	36.2 (32.5-40.0)	248 (6)
CO	364	P	0.5	1.71 (0.97-3.00)	0.18 (0.13-0.25)	2.9 (4)
		L		7.94 (3.44-18.30)	0.18 (0.15-0.21)	2.2 (4)

[a] Data for generations 30 (C30) and 31 (C31) of *B. dorsalis* eggs exposed to 2.8°C. Units are days. Data provided by A. K. Burditt, Jr, USDA, ARS, retired.

[b] Data (Mango) from time-temperature experiments on wild and laboratory strains of *C. capitata* in mangoes immersed in water at 46.1 ± 0.25°C (Sharp & Picho-Martinez 1990). Units are in minutes.

[c] Data for mexacarbate topically applied to western spruce budworm (CO).

[d] n is the total number of test subjects tested in the experiment.

[e] P is the probit model, L is the logit model.

[f] Max is the largest dose or time tested.

Number of Doses

Müller & Schmitt (1990) addressed the problem of how to choose the number of doses to provide the best estimate of the LD_{50} based on the criterion of minimizing the asymptotic variance at the 50% response level. They concluded that use of as many doses as possible is preferable and that use of only three doses cannot be recommended. For any routine bioassay, use of at least five doses has been recommended. Far more are probably required to obtain a reliable estimate of 99.9968% mortality; most would have to cause 100% mortality. Regardless of the number of doses tested, a line cannot be estimated if all but one cause 100% mortality (Robertson & Preisler 1992).

Goodness-of-fit

The measure of how well data fit the assumptions of the model is the goodness-of-fit. The usual way to test fit is with a chi-square test. In a chi-square test, values (responses) predicted by the model are compared with values actually observed in the bioassay. If the values differ significantly at $P = 0.05$, the model does not fit the data and a more appropriate model should be sought. However, alternatives to the probit or logit models are not easy to use (Robertson & Preisler 1992).

Various methods to compensate for lack-of-fit have been used. Finney (1971) suggests multiplying the variances by a heterogeneity factor to account for extra variation that causes poor fit. A similar method is used in various probit or logit programs (e.g., POLO-PC [LeOra Software Inc. 1987], SAS PROBIT [SAS Institute Inc. 1982]). However, this method does not identify the reasons why the data do not fit the model. A more meaningful approach is to examine possible causes of lack-of-fit by means of residual plots (Preisler 1988).

A residual is the difference between an observed response and an expected response. Because each binomial response has a variance related to the size of the response, it must be standardized by division by its standard error before residuals are plotted. The standard error is

$$\sqrt{np(1-p)} \tag{3}$$

where n is the number of subjects tested at the specific dose. Estimated response probability is designated as p. For good fit, residuals plotted against predicted values or against dose usually lie within a horizontal band around zero (mostly 0 ± 2).

Preisler (1988) listed five causes of significant departure from the probit-binomial model that might cause lack-of-fit; some may be identified by residual plots. First, outliers that do not fit the model might be the cause of lack-of-fit, especially if they occur at a response level near 0 or 100%. An outlier should be discarded from the data set if strong evidence suggests that an error in recording has occurred. Otherwise, the outlier could indicate the existence of an important independent variable. A second cause of lack-of-fit may be omission of a significant explanatory variable (e.g., body weight, temperature) from the model. For example, a univariate model has been used but a multiple probit or logit model is appropriate. A third cause of lack-of-fit is that the probit curve might not fit the data. Fourth, responses of subjects with a factor in common might be correlated; as a result, error terms are not independent. For example, insects from the same generation might be more alike in their response to a pesticide than they are compared with insects in another generation. Finally, error terms may not be binomial (Preisler 1988). For this cause of lack-of-fit, Preisler (1988) described a method to obtain maximum likelihood estimates for parameters in a compound-binomial-probit model. This method is based on addition of a random effect factor to the model.

Experimental Designs When Time Is a Variable

In commodity treatment bioassays, time-dose-response data have frequently been analyzed by fitting probit lines (with time replacing dose) to mortality data for a fixed dose over time. Unless different groups are used for each observation period, this method is not appropriate because responses at different time points will be correlated regardless of the model used (Preisler 1988).

Bioassays that include both time and dose as variables can have one of two possible designs. With the independent sampling design, separate groups of test subjects are treated with a fixed dose. Each group is then observed for a different period of time, and numbers of responses are recorded at each observation period. For the probit or logit models to be appropriate, responses that occur in a given group of test subjects must be recorded only once.

The other alternative, the serial sampling design, involves treatment of each group of subjects with a given dose, inclusion of several doses in the experiment as a whole, and recording of responses for each dose group at a series of times after treatment. A number of observation intervals are necessary because responses differ with each dose (i.e., a high dose might kill more test subjects than a low dose in a given period of time, whereas a low dose might kill fewer subjects during the same interval). Thus, a serial design involves both treatment with a series of doses and a series of observations within each treatment group.

General Statistical Models for Studies with Time

The probability of response by a given time t can be modeled by the probit, logit, or other curves with time as the single independent variable. A general statistical model for the binary response with time as the explanatory variable is

$$p_i = F(\alpha + \beta t_i) \tag{4}$$

where p_i is the probability of response, t_i is the ith time, α is the intercept of the regression line, β is the slope of the regression line, and F is a distribution function (such as probit, logit, or complementary log-log [CLL]).

The general statement of the CLL relationship in the serial time-mortality design is as follows: For a fixed exposure time t_j ($j = 1,...,J$), to a treatment at a concentration d_i ($i = 1,...,I$), probability of mortality of a test subject by time t_j is

$$p_{ij} = 1 - \exp[-\exp(\gamma_j + \beta \log_{10}(d_i))] \tag{5}$$

where β is an unknown parameter and γ_j are unknown categorical variables corresponding with the times t_j. The linear part of equation (5), $\gamma_j + \log_{10}(d_i)$, is called the CLL line. The model assumes that $\log_e(-\log_e(1 - p_{ij})$ is linear in the covariates.

For the independent time-mortality design, γ_j is replaced by the logarithm of t_j and $\log_{10}(d_i)$ is replaced by the constant $\log_{10}(D)$ where D is the fixed dose used in the experiment. This statement of the CLL relationship is actually the Weibull function, which has been used to model responses to some chemicals over time (Su et al. 1987). Further details of analyses of studies with time as a variable are provided by Preisler & Robertson (1989) and Robertson & Preisler (1992).

Examples of Problems

Data from three bioassays, including two commodity treatment experiments done to estimate an $LD_{99.9968}$ for oriental fruit fly, *Bactrocera dorsalis* (Hendel), and Mediterranean fruit fly, *Ceratitis capitata* (Wiedemann), and sample data from a pesticide bioassay with western spruce budworm, *Choristoneura occidentalis* Freeman, done to estimate an LD_{50} and LD_{90} (Savin et al. 1977) demonstrate the special problems inherent in use of the probit analysis, especially at the probit 9 level.

Table 4.1 shows that the choice of model for each of three bioassay data sets had a negligible effect at LT or LD_{90}, but a large effect on estimates for LT or $LD_{99.9968}$. For the studies with oriental fruit fly and Mediterranean fruit fly, large values of the chi-square goodness-of-fit statistic versus degrees of freedom indicated that neither model adequately fits the data. The goodness-of-fit for the western spruce budworm data is very good for both models; however, estimates of the $LD_{99.9968}$ for the two models were dramatically (and significantly) different. Estimation of the $LD_{99.9968}$ required extrapolation of the curve to a dose that was far outside the range tested in the experiment. Extrapolation is not a sound practice; data such as those for western spruce budworm should only be used to estimate a maximum of an LD_{95} because this was the highest kill actually observed.

As mentioned previously, large chi-square values such as those for oriental fruit fly and Mediterranean fruit fly data sets are usually managed by multiplication of the variances and covariances by the heterogeneity factor (SAS Institute, Inc. 1982, Sharp & Picho-Martinez 1990). Besides ignoring the fundamental problem, this practice assumes that all causes of poor fit affect calculations of the standard error but not the point estimates. Causes of poor fit can be studied by examining plots of standardized residuals for the dose levels tested (Preisler 1988). When we examined plots of the temperature data for two generations of oriental fruit fly, no systematic patterns that might indicate departure from the probit model were apparent (Fig. 4.1A, B). Both plots showed some evidence of extra binomial variation (i.e., variation about the mean is greater than expected under the binomial model); an outlier was apparent in the data for generation 31 of oriental fruit fly exposed to 2.8°C. However, the outlier did not appear to be the only cause of bad fit because chi-square was still large when the outlier was removed. A model that incorporates extra binomial variation within the probit framework might give a better fit. This was done by adding a random effect factor to the probit line as follows:

$$y_i = \alpha + \beta \log_{10}(d_i)[\text{or } (t_i)] + \sigma z_i \qquad (6)$$

where y is the probit of response, d_i is the dose (or t_i is time), z is an unobserved standard normal variate, and where (α, β, σ) are the unknown parameters to be estimated. The parameter σ is a measure of the extra variation in the data. A maximum likelihood procedure for estimating the parameters in equation (6) is outlined by Preisler (1988). For generation 31, the outlier was not included. The chi-square values thus obtained seem to indicate that the data fit this model reasonably

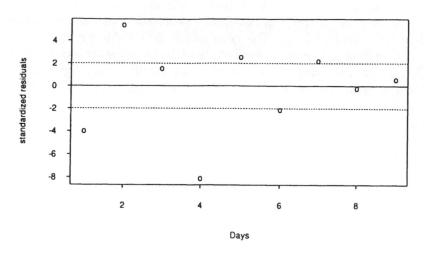

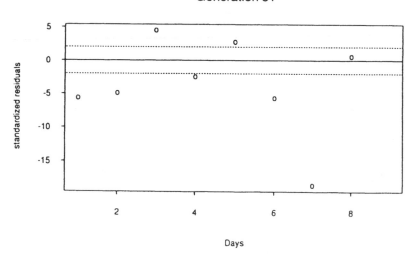

FIGURE 4.1. Plots of standardized residuals developed by fitting probit curves to responses of *B. dorsalis* eggs from generation 30 (A) and 31 (B) exposed to 2.8°C for different numbers of days. Most of the residuals are outside the horizontal band of ± 2 units about zero.

TABLE 4.2. Results for seven generations of oriental fruit fly stored at 2.8°C, with estimates obtained by fitting a probit model with a random effect factor in the probit line to account for extra binomial variation[a]

Generation	Longest Treatment, d	$LT_{99.9968}$ (95% CL), d	Chi-square	df
30	9	9.52 (8.79-10.24)	2.86	6
31	7	10.64 (8.47-12.82)	1.52	4
32	10	9.46 (8.71-10.21)	8.96	7
33	8	9.38 (8.46-10.30)	2.54	5
34	9	9.30 (7.89-10.70)	1.69	6
35	9	10.10 (9.17-11.04)	3.12	6
37	12	10.54 (10.34-10.75)	7.25	9

[a] Data from A. K. Burditt, Jr., USDA, ARS, retired.

reasonably well. All estimates of $LT_{99.9968}$ are larger than those produced by the simple probit model. Estimates of $LT_{99.9968}$ for the various generations also show differences in response (Table 4.2). For example, the 95% confidence limits for generations 30 and 36 do not overlap. The average $LT_{99.9968}$ over the seven generations in Table 4.2 is 9.855 d (SD = 0.56). Assuming that the seven generations included in this experiment provide a random sample of the responses of present and future generations of oriental fruit fly, 95% of the generations would have a $LT_{99.9968}$ that is less than 9.855 + (1.65 x 0.56) = 10.78 d. Therefore, the treatment level that could be used in a confirmatory test to verify that the treatment meets quarantine security requirements is 10.78 d when eggs are exposed to a temperature of 2.8°C.

One reason that the probit model does not fit the data for Mediterranean fruit fly in mangoes is that mortalities were estimated rather than known (Sharp & Picho-Martinez 1990). In these experiments, the number of larvae treated at each level was estimated from a control group not immersed in water. Both controls and treated mangoes had been exposed to natural oviposition by adult fruit flies. Unfortunately, infestation by natural oviposition can be variable (Armstrong 1983, Armstrong & Couey 1984).

The problem of n being unknown causes special problems in the analysis when mortality is calculated to be negative. Usually, data points with negative mortality are disregarded (Sharp & Picho-Martinez 1990). A more efficient way to deal with experiments of this type is to model the observed survival rather than death by using the Poisson or Poisson-lognormal models (Preisler & Robertson 1992). Consideration of survival not only avoids the problem of apparent negative mortality, it includes the extra source of variation caused by the fact that exact numbers of larvae treated at each level are unknown.

Confirmatory Tests

In confirmatory tests, large numbers of the target species are exposed to a fixed treatment level (usually the LD or $LT_{99.9968}$); based on the number surviving, an upper 95% boundary for the probability of at least one survivor is calculated. For example, if n larvae are treated and no survivors are observed, then the 95% upper boundary is $p_u = 1 - (0.05)^{1/n}$. If $n = 95,000$, for instance, then $p_u = 31.53$ per million. Because the probit 9 requirement is <32 survivors per million, the treatment can be said to meet the probit 9 criterion. If one or more survivors is observed, then the tables given by Couey & Chew (1986) can be used to calculate p_u. At the stage of confirmatory testing, the assumption that the data follow a probit curve is unnecessary.

Infestation Level and Lot Size

In the previous sections, we have described experimental and statistical procedures used to determine a treatment level for use in quarantine treatments, and to estimate the efficacy of a given treatment by a confirmatory test. The efficacy of a treatment is the percentage of eggs, larvae, or adults that a given treatment is expected to kill. Because efficacy is a percentage, it does not provide an estimate of the actual number of survivors (or those killed) in a given lot of fruit. The actual number depends on the proportion of infested fruit, the mean infestation level of that infested fruit, and the size of the lot. With these factors in mind, Landolt et al. (1984) proposed "that the level of security recommended for quarantine treatments of tephritid infested fruit be set at a particular probability of an introduction occurring, specified as the probability of a potential mating pair arriving per shipment."

Probability of Survivors per Lot

By making some assumptions, for example, about the distribution of the number of larvae per fruit, we can derive a formula for the probability of the number of live larvae in a shipment of a particular size. First, we assume that the emergence of two adults of the opposite sex is the minimum required to establish a population. Then, using a

Poisson approximation (Baker et al. 1990), we calculate

$$Pr \text{ (}r \text{ survivors in a lot of size } N) = \frac{e^{-\lambda}\lambda^r}{r!} \tag{7}$$

where $\lambda = N\phi\mu p$, N = lot size, ϕ = treatment efficacy, μ = mean number of larvae per infested fruit, and p = proportion of fruit infested. Therefore, as given in Landolt et al. (1984), the probability of having ≥ 2 larvae that will emerge as adults of the opposite sex in a lot size N is:

$$q = Pr[r \geq 2] = [1 - e^{-\lambda/2}]^2 \tag{8}$$

For example, with a lot size of N = 60,000 fruit, a proportion of infested fruit p = 0.001 (one infested fruit per 1000), a mean number of larvae per infested fruit of μ = 10 and a treatment efficacy of $\phi = 32/10^6$, λ = 0.0192 and q = 0.00009128. Therefore, the probability that any lot will have ≥ 2 survivors is one in 10,955. If the proportion of infested fruit p equals 0.0001 (i.e., one infested fruit per 10,000 fruit), then the expected probability of ≥ 2 survivors is one in 1,085,776.

These calculations indicate that the lot size will influence the risk of introducing a mating pair from a shipment. For practical purposes, a maximum lot size arriving at a given point of entry at a given time could thus be set to minimize the probability of introducing a mating pair into a country.

Confidence Bounds for Probability of Survivors

The probability q in equation (8) is only an estimate because the efficacy of treatment, the proportion of infested fruit, and the mean infestation level are estimates that include their own error terms (Frampton et al. in preparation). Given the standard errors of the estimates $\hat{\phi}$, $\hat{\mu}$, and $\hat{p}$, the delta-method for calculating approximate standard errors of functions of random variables could be used to calculate a standard error and a 95% confidence for $\hat{q}$ (Robertson & Preisler 1992). Alternatively, an upper boundary for $\hat{q}$ could be calculated given similar bounds for the efficacy of a treatment, the proportion of infested fruit, and the infestation level of fruits from a given area.

Maximum Pest Limits and Acceptance Sampling

New Zealand was the first country to attempt to define and

implement biologically sound criteria for pest exclusion. The maximum pest limit (Baker et al. 1990) recognizes the validity of the suggestion made by Landolt et al. (1984) that the probability of successful reproduction is the basis for establishment of an exotic pest species in an area where it does not occur.

Baker et al. (1990) described methods for calculating the number of fruits to be sampled in order to detect, with some accuracy, the maximum allowable infestation level prior to treatment of a known efficacy. Their method relies on four assumptions: (1) the treatment acts independently on different fruit fly individuals; (2) the mean number of fruit flies within an infested fruit is known; (3) the efficacy of the treatment to be used in the country of origin is known (it is not necessary for the efficacy to be probit 9); and (4) the maximum lot size assembled per day at one location is known.

Further research is needed to refine the concept of the maximum pest limit. For example, another determinant for pest establishment is the climatic suitability of the new environment into which the arthropod might be carried in an infested commodity. In the future, the concepts of maximum pest limits and ecoclimatic indices (Worner 1988) might be integrated to provide even more refined pest exclusion criteria.

Conclusions

Results for the three sample data analyses described in the section concerning laboratory testing suggest that three criteria should be used by regulatory agencies to ensure the statistical validity of bioassays of possible treatments for commodity protection. First, the experimental design must be such that it covers the lethal dose of likely interest. Estimates obtained by extrapolation outside the range of doses actually tested are always suspect. Implicit in the experimental design are use of adequate numbers of test subjects and adequate replications of the experiment. Second, the model used must adequately fit the data. If fit is not adequate, causes for poor fit should be identified. If the model cannot be adapted, another model should be sought. Finally, commodity treatment bioassays should be done with more than one generation of the target species to account for natural variation among generations and experiments.

Requirements of scientific publications can help to ensure that the statistical analyses used in commodity treatment bioassays are scientifically based. The requirement of a regulatory agency for a probit 9 estimate should not mean that meaningless values can be presented in the scientific literature. An author attempting to publish probit 9 values should be required to present evidence that the first two (and preferably, all three) considerations described in the preceding paragraph have been taken into account. The interpretation of bioassay data presented by quarantine entomologists must be statistically valid regardless of bureaucratic requirements.

Finally, we recommend use of the more general (and less restrictive) terms (LD_x, LT_x, LC_x) to replace probit 9 requirement. When the distribution function used in the binomial model is the normal distribution, an $LD_{99.9968}$ would be equal to the probit 9. However, different distribution functions could be used as well, permitting investigators to chose the model that best fits their data.

References

Armstrong, J. W. 1983. Infestation biology of three fruit fly (Diptera: Tephritidae) species on 'Brazilian,' 'Valery,' and 'William's' cultivars of banana in Hawaii. J. Econ. Entomol. 76: 539-543.

_____. 1985. "Pest Organism Response to Potential Quarantine Treatments," in *Proceedings, Regional Conference on Plant Quarantine Support for Agricultural Development*. Pp. 25-31. Association of South East Asian Nations, Plant Quarantine Centre and Training Institute, Serdang, Selangor, Malaysia.

Armstrong, J. W. & H. M. Couey. 1984. Methyl bromide treatments at 30°C for California stonefruits infested with Mediterranean fruit fly (Diptera: Tephritidae). J. Econ. Entomol. 77: 1229-1232.

_____. 1989. "Fumigation, Heat and Cold," in A. S. Robinson & G. Hooper, eds., *World Crop Pests, Vol. 3B, Fruit Flies. Their Biology, Natural Enemies and Control*. Pp. 411-424. Amsterdam: Elsevier.

Armstrong, J. W., J. D. Hansen, B.K.S. Hu & S. A. Brown. 1989. High-temperature, forced-air quarantine treatment for papayas infested with Tephritid fruit flies (Diptera: Tephritidae). J. Econ. Entomol. 82: 1667-1674.

Baker, A. C. 1939. The basis for treatment of products where fruitflies are involved as a condition for entry into the United States. USDA Circular 551.

Baker, R. T., J. W. Cowley, D. S. Harte & E. R. Frampton. 1990. Development of a maximum pest limit for fruit flies (Diptera: Tephritidae) in produce imported from New Zealand. J. Econ. Entomol. 83: 13-17.

Berkson, J. 1951. Why I prefer logits to probits. Biometrics 33: 327-339.

Brown, B. W., Jr. 1966. Planning a quantal assay of potency. Biometrics 22: 322-329.

Burditt, A. K., Jr., & F. P. Hungate. 1988. Gamma irradiation as a quarantine treatment for cherries infested by western cherry fruit fly (Diptera: Tephritidae). J. Econ. Entomol. 81: 859-862.

Copenhaver, T. W. & P. W. Mielke. 1977. Quantit analysis: a quantal assay refinement. Biometrics 33: 175-186.

Couey, H. M. & V. Chew. 1986. Confidence limits and sample size in quarantine entomology. J. Econ. Entomol. 79: 887-890.

Finney, D. J. 1964. *Statistical Method in Biological Assay, 2nd Ed.* London: Griffin.

_____. 1971. *Probit Analysis. 3rd. Ed.* Cambridge University Press.

Frampton, C. M., R. J. Ivess & E. R. Frampton. In preparation. A statistical approach for comparing the efficacy of area freedom, non-host status and disinfestation treatments as "treatments" against fruit flies (Diptera: Tephritidae).

Freeman, P. B. 1970. Optimal Bayessian sequential estimation of the median effective dose. Biometrika 57: 79-89.

Gaffney, J. J. & J. W. Armstrong. 1990. High-temperature forced-air research facility for heating fruits for insect quarantine treatments. J. Econ. Entomol. 83: 1959-1964.

Gaffney, J. J., G. J. Hallman & J. L. Sharp. 1990. Vapor heat research unit for insect quarantine treatments. J. Econ. Entomol. 83: 1965-1971.

Haverty, M. I. & J. L. Robertson. 1982. Laboratory bioassays for electing candidate insecticides and application rates for field tests on the western spruce budworm. J. Econ. Entomol. 75: 183-187.

Landolt, P. J., D. L. Chambers & V. Chew. 1984. Alternative to the use of probit 9 mortality as a criterion for quarantine treatments of fruit fly (Diptera: Tephritidae)-infested fruit. J. Econ. Entomol. 77: 285-287.

LeOra Software, Inc. 1987. POLO-PC: A user's guide to Probit Or LOgit analysis. LeOra Software, Inc., 1119 Shattuck Ave., Berkeley, California.

Monro, H.A.U. 1969. Manual of fumigation for insect control. FAO Agriculture Studies No. 79. Rome, Italy.

Müller, H-G. & T. Schmitt. 1990. Choice of number of doses for maximum likelihood estimation of the ED50 for quantal dose-response data. Biometrics 46: 117-126.

Ottoboni, M. A. 1984. *The Dose Makes the Poison.* Berkeley, California: Vincente Books.

Podgwaite, J. D., R. C. Reardon, D. M. Kolodny-Hirsh & G. S. Walton. 1991. Efficacy of ground application of the gypsy moth (Lepidoptera: Lymantriidae) nucleopolyhedrosis virus product, Gypchek. J. Econ. Entomol. 84: 440-444.

Preisler, H. K. 1988. Assessing insecticide bioassay data with extra-binomial variation. J. Econ. Entomol. 81: 759-765.

Preisler, H. K. & J. L. Robertson. 1989. Analysis of time-dose-mortality data. J. Econ. Entomol. 82: 1534-1542.

_____. 1992. Estimation of treatment efficacy when the number of test subjects is unknown. J. Econ. Entomol. 85: 1033-1040.

Robertson, J. L. & H. K. Preisler. 1992. *Pesticide Bioassays with Arthropods*. Boca Raton, Florida: CRC Press.

Robertson, J. L., K. C. Smith, N. E. Savin & R. J. Lavigne. 1984. Effects of dose selection and sample size on the precision of lethal dose estimates in dose-mortality regression. J. Econ. Entomol. 77: 833-837.

SAS Institute, Inc. 1982. User's guide: statistics. SAS Institute, Cary, North Carolina.

Savin, N. E., J. L. Robertson & R. M. Russell. 1977. A critical evalution of bioassay in insecticide research: likelihood ratio tests of dose-mortality regression. Bul. Entomol. Soc. Amer. 23: 257-266.

Sharp, J. L. & H. Picho-Martinez. 1990. Hot-water quarantine treatment to control fruit flies in mangoes imported into the United States from Peru. J. Econ. Entomol. 83: 1940-1943.

Sharp, J. L., J. J. Gaffney, J. I. Moss & W. P. Gould. 1991. Hot-air treatment device for quarantine research J. Econ. Entomol. 84: 520-527.

Spitler, G. H. & H. M. Couey. 1983. Methyl bromide fumigation treatments of fruits infested by the Mediterranean fruit fly. J. Econ. Entomol. 76: 547-550.

Su, N-Y., M. Tamashiro & M. I. Haverty. 1987. Characterization of slow-acting insecticides for remedial control of the Formosan subterranean termite (Isoptera: Rhinotermitidae). J. Econ. Entomol. 80: 1-4.

Tsutakawa, R. K. 1980. Selection of dose levels for estimating a percentage point of a logistic quantal response curve. Applied Statistics 29: 25-33.

Weissling, T. J. & L. J. Meinke. 1991. Potential of starch encapsulated semiochemical-insecticide formulations for adult corn rootworm (Coleoptera: Chrysomelidae) control. J. Econ. Entomol. 84: 601-609.

Worner, S. P. 1988. Ecoclimatic assessment of potential establishment of exotic pests. J. Econ. Entomol. 81: 973-983.

5

Fumigation

Victoria Y. Yokoyama

Fumigation with chemical fumigants is the primary quarantine treatment method. Ware (1978) described fumigants as small, volatile, organic molecules, commonly with one or more halogens, that become gases at temperatures >4.4°C. Fumigants are effective chemical control agents because most are highly toxic, economical to use, easy to apply, and diffuse rapidly. Such advantages have been recently outweighed by their potential detrimental effects on human health and the environment. The fumigant, ethylene dibromide, which was previously used for quarantine treatments was banned in 1984 by the United States Environmental Protection Agency because it was found to cause cancer in laboratory animals (Anonymous 1984). Methyl bromide replaced ethylene dibromide as a quarantine treatment, but the future use of methyl bromide also has been jeopardized because it may deplete atmospheric ozone (Anonymous 1992a). Many chemical compounds can be used as fumigants. Only methyl bromide is widely used in quarantine treatments.

The Manual of Fumigation for Insect Control (Bond 1984) is recommended highly as a reference for fumigant applications. The manual contains information on procedures and is available from Bernan Unipub, 4611F Assembly Drive, Lanham, MD 20706, USA, Telephone (301) 459-7666. The manufacturer of each fumigant is a good source of specific application information and material safety data sheets. Additional information regarding pest control, commodity response, and the chemistry and toxicology of each fumigant can be obtained through independent literature reviews.

This chapter will help orient the reader to the use of fumigation in quarantine treatments. The chapter presents fumigation treatments in use, fumigation treatments under development, studies on fumigant residues, and research methods to develop fumigation as a single quarantine treatment to control arthropod pests of regulatory concern in exported or imported agricultural commodities.

Certified Fumigation Treatments

The United States Department of Agriculture, Animal and Plant Health Inspection Service, Plant Protection and Quarantine (USDA-APHIS-PPQ) Treatment Manual describes quarantine treatments for commodities primarily imported into the United States of America (U.S.) (Anonymous 1992b). Fumigation procedures that have been approved as quarantine treatments to control arthropods of regulatory concern in foods are published in this manual. The manual can be obtained from the USDA-APHIS-PPQ, Professional Development Center, 7340 Executive Way, Suite A, Frederick, MD 21701, USA, Telephone (301) 663-0342. The manual contains sections on fumigants, treatment equipment and accessories, treatment procedures, facilities, schedules, emergency aid and safety, and is highly recommended as a reference. Table 5.1 summarizes approved fumigation schedules that are listed in the manual. The manual is subject to constant revision. Inquiries concerning the content or use of the manual should be directed to the USDA, APHIS, Hoboken Methods Development Center, 209 River Street, Hoboken, NJ 07030, USA, Telephone (201) 659-9099.

The Export Certification Manual (Anonymous 1986) is published and used by the USDA-APHIS-PPQ as a reference for export quarantine requirements for each country to which a commodity is shipped. Table 5.2 summarizes approved fumigation schedules that are listed in the Export Certification Manual which is also subject to constant revision. Inquiries concerning exports should be directed to the USDA-APHIS-PPQ, Export Certification Unit, 6505 Belcrest Road, Federal Building, Hyattsville, MD 20782, USA, Telephone (301) 436-8537. Export summaries from the manual for each country of concern can be obtained from the USDA-APHIS-PPQ, Professional Development Center, Room G-195, Federal Building, 6505 Belcrest Road, Hyattsville, MD 20782, USA, Telephone (301) 436-4478.

Further information regarding quarantine treatments can be obtained from the Department of Agriculture of each state. References such as The State of California Commodity Treatment Manual (Fiskaali 1989) complement the USDA-APHIS-PPQ Treatment Manual. In California, the County Agriculture Department participates with the California Department of Food and Agriculture and USDA-APHIS-PPQ to administer quarantine treatments and can be contacted for information regarding the export and import of commodities in each county.

Reports on efficacy testing of the methyl bromide fumigation schedules in the USDA-APHIS-PPQ Treatment Manual and the Export Certification Manual can be found in the literature if the treatment was recently developed. Test results of fumigation schedules that were published in the manual at an earlier date may not be available. For example, methyl bromide fumigation schedules published in the Export Certification Manual were tested on the immature stages of oriental fruit moth, *Grapholita molesta* (Busck), to verify efficacy of the treatment to control the pest in nectarines (Yokoyama et al. 1987a) because earlier test data were not available.

Fumigation Treatments Under Development

Quarantine treatments that use fumigation as a single treatment to control arthropods of regulatory concern are developed from efficacy testing data. Tests to demonstrate that a fumigation schedule can control the pest(s) of concern are the basis of the quarantine treatment and additional testing may be necessary to show that selected fumigation schedules will not cause commodity damage. The proposed quarantine treatments are intended to rescind suspensions or bans on commodities in areas where the pests of concern are not found. Treatments that have been developed but not yet certified are shown in Table 5.3. Acceptance of proposed quarantine treatments by export and import regulatory agencies will provide continuous shipments and new markets for agricultural commodities.

Greater emphasis recently has been placed on the development of alternative quarantine treatments to chemical fumigants because of the uncertain future for fumigants. Many pesticide bans have resulted from evaluation of human health risks and environmental problems associated with their use.

Phytotoxicity

Commodity damage (Chapter 17) caused by exposure to fumigants has been described as phytotoxic effects. Such effects may impair the appearance of the commodity, cause off-flavors or odors, or shorten the shelf-life of the product.

The dose, duration and time of exposure, and temperature that cause commodity damage must be determined in basic tests that are similar to those to determine 100% mortality of the target pest. Fumigation conditions that cause commodity damage must be identified and the types of damage must be described. The effect of fumigation on quality must be evaluated over a period of time that simulates shipping and retail conditions.

Residues

Quarantine treatments that utilize chemical fumigants must not leave undesirable residues in the commodity or liberate gas concentrations from packaging materials after treatment. The amount of fumigant residues in treated food is affected by the type and concentration of fumigant, the nature of the commodity, and exposure temperature and duration. Treatment of foods with methyl bromide results in a small portion of the fumigant persisting in the commodity, a small portion reacting with the commodity to form relatively innocuous inorganic bromide residues, and the remaining fumigant desorbing from the product (Sell & Moffitt 1990).

Table 5.4 shows references for residue analysis in commodities for which methyl bromide fumigation has been developed as a quarantine treatment. Information concerning fumigant residues and tolerances allowed in foods that have been subjected to quarantine treatment can be obtained from the USDA-APHIS-PPQ, Hoboken Methods Development Center, 209 River Street, Hoboken, NJ 07030, USA, Telephone (201) 659-9099.

Methods to Develop a Fumigation Treatment

Fumigations developed as quarantine treatments are investigated in two stages: basic tests and large-scale efficacy tests. The basic tests should be performed in a series of small experimental fumigation chambers. Small fumigation chambers such as 28 liters will greatly expedite the research because several doses can be tested at one time. Establishing a practical fumigation temperature and duration before the basic tests will eliminate the temperature and time variables that increase testing requirements.

Dose-response relationships in basic tests are determined by testing a series of doses against a life stage at a specific temperature and time to determine the LD_{50} (Lethal dose to 50% of the test population). The LD_{50} values are compared among life stages of the same species to determine the least susceptible stage.

The minimum dose to cause 100% mortality is determined for each life stage of each species in the basic tests. Results of the dose-response tests are used to establish the dose to be used in the quarantine treatment. This dose will control the least susceptible stage of each species and may be higher than the dose to cause 100% mortality if commodity damage does not occur.

Life stages tested are stages that may occur in the commodity. Numbers of insects tested in basic tests require three replicates of a minimum of five doses per life stage. A substantial number of insects needs to be tested per dose per replicate per species. The number of insects that must be tested in basic tests is influenced by the rearing method and the capacity to produce the insect in high numbers.

Controls must be conducted with each test, and test mortality must be corrected for control mortality. The period to evaluate mortality and the method to determine mortality after treatment must be appropriate for the species and life stage tested. Methods to estimate the test population in the infested commodity that was fumigated must be acceptable to regulatory agencies if actual counts of treated individuals are not feasible.

Data in basic tests can be analyzed by the probit procedure using statistical software such as Statistical Analysis Systems (SAS 1988) or the POLO-PC (LeOra Software 1987) program in order to determine the LD_{50} estimates. Pesticide bioassay methods have been revised by Robertson & Priesler (1992).

Large-scale tests are used to confirm the fumigation schedule selected for use as a quarantine treatment to control the target pest based on the results of basic tests. The fumigation schedule includes dose, temperature, duration, and chamber load. Large-scale tests are used to confirm the efficacy of the quarantine treatment under commercial conditions.

The total number of insects and replicates tested in large-scale tests must be acceptable to regulatory agencies and must fulfill quarantine security requirements. The commodity is infested with the least susceptible stage to fumigation and fumigated in shipping containers of the type that will be exported to or from the U.S. Controls must be used for each replicate of the large-scale test. The number of insects used in each test replicate must be corrected by the amount of mortality in the controls.

Gas concentrations and fruit pulp temperatures are reported for at least three locations in the chambers during fumigation. The concentration multiplied by time (c x t) product is a useful value to report the actual concentration of fumigant in the chamber during each test. The c x t product is especially useful for methyl bromide fumigations. Fumigant residues in the commodity are determined immediately after fumigation and intervals after treatment.

The development of a quarantine treatment to control codling moth, *Cydia pomonella* (L.), in nectarines for export to Japan where the insect is not found, is an example of the research needed to fulfill regulatory agency requirements to ship a commodity from the U.S. to a foreign country when the commodity may be infested with a pest of concern. Basic tests were used to demonstrate doses required to control eggs 1 d old, the stage least susceptible to methyl bromide (Gaunce et al. 1980), that may occur on nectarines. The results were used to establish a quarantine treatment schedule of 48 g per cubic meter of methyl bromide for 2 h at 21°C or above (Yokoyama et al. 1987b). Methods to rear codling moth in high numbers were developed and large-scale tests confirmed the efficacy of the quarantine treatment to control codling moth under commercial conditions in field bins (Yokoyama 1988). Studies of the effect of methyl bromide fumigation on fruit quality showed no damage to the cultivars proposed for shipment to Japan and fumigant residues were below the U.S. tolerance (Harvey et al. 1989).

Nectarines fumigated in field bins and then packed after treatment were first shipped to Japan in 1988 after regulatory agencies approved the research results. Further research to modify the technology of the quarantine treatment for nectarines exported to Japan included efficacy testing to add more fruit cultivars to the list of cultivars approved for export (Yokoyama et al. 1990a) and to fumigate fruit in shipping containers rather than field bins prior to export (Yokoyama et al. 1990b, 1993).

A quarantine treatment for in-shell walnuts was also developed in a similar manner. Results of large-scale tests showed that a methyl bromide fumigation under reduced pressure would control codling moth to a level that would maintain quarantine security in in-shell walnuts (Hartsell et al. 1991a) and fumigant residues were below U.S. tolerances (Hartsell et al. 1991b).

Acknowledgments

The author is grateful to Gina T. Miller, Preston L. Hartsell (USDA, ARS, Horticultural Crops Research Laboratory, Fresno, CA), and Robert Berninger (USDA-APHIS-PPQ, Hoboken Methods Development Center, Hoboken, NJ) for their assistance with the preparation of this chapter.

TABLE 5.1. Commodities with pests of regulatory concern in the Plant Protection and Quarantine Treatment Manual (Anonymous 1989) that require a mandatory methyl bromide fumigation as a single quarantine treatment when imported to or exported from the USA

Commodity	Destination	Pests of Regulatory Concern	Treatment Schedule
DECIDUOUS FRUITS			
Apple Cherry Pear	Chile to USA	Found on Inspection	T101
Apple Pear	New Zealand to USA	Tortricidae, leafroller, and olethreutine moths	T101
Cherry	USA to Japan	*Cydia pomonella* (L.), codling moth; *Rhagoletis indifferens* Curran, western cherry fruit fly	T101
Grape	USA imports	*Lobesia botrana* (Schiffermueller), vine moth; *Ceratitis capitata* (Wiedemann), Mediterranean fruit fly	T101
Kiwi *Opuntia* cactus fruits	USA imports	*Ceratitis capitata* (Wiedemann), Mediterranean fruit fly	T101
Pear	Australia to USA	Tortricidae, leafroller, and olethreutine moths	T101
TROPICAL AND SUBTROPICAL FRUITS			
Avocado	Hawaii to mainland USA	*Ceratitis capitata* (Wiedemann), Mediterranean fruit fly; *Bactrocera cucurbitae* (Coquillett), melon fly; *Bactrocera dorsalis* (Hendel), oriental fruit fly; *Hemiberlesia lataniae* (Signoret), latania scale	T102

(Continues)

TABLE 5.1. (Continued)

Commodity	Destination	Pests of Regulatory Concern	Treatment Schedule
Citrus	USA imports	Aleurocanthus woglumi Ashby, citrus blackfly	T102
Ethrog (citrus)	USA imports	Prays citri Miller, citrus flower moth	T102
Grapefruit Orange Tangerine	Mexico to USA	Anastrepha spp., fruit flies	T102
Pineapple	USA imports	Found on inspection	T102
NUTS			
Almond with husk Walnut with husk	USA imports	Ceratitis capitata (Wiedemann), Mediterranean fruit fly	T103
Chestnut	USA imports except Canada & Mexico	Cydia spendana (Hübner); Curculio spp., acorn and nut weevils	T103
Macadamia	USA imports	Cryptophlebia illepida (Butler), koa seedworm	T103
VEGETABLES			
Asparagus	Australia or New Zealand to USA; waived with phytosanitary certificate	Halotydeus destructor (Tucker), red-legged earth mite	T104
Beans Lentils	USA imports	Found on inspection	T104[a]

Commodity	Origin/Condition	Pest / Inspection	Treatment
Cipollini (onion)	USA imports	*Exosoma lusitanica* (L.)	T104
Faba beans	USA imports		T104[b]
Garlic	Italy and Spain to USA; waived with phytosanitary certificate	Found on inspection	T104
Green corn on the cob	USA imports	Found on inspection	T104
Green pod vegetables	USA imports	*Maruca testulalis* (Geyer), bean pod borer; *Epinotia aporema* (Walsingham), tortricid moth; *Cydia fabivora* (Meyrick), tortricid moth	T101,T104
Horseradish	Hawaii imports	*Baris lepidii* Germer, imported crucifer weevil	T104
Horseradish	USA imports except Hawaii	Found on inspection	T104
Leafy vegetables	USA imports	Found on inspection	T104
Onion	USA imports	Found on inspection	T104
Okra	USA imports and exports	*Pectinophora gossypiella*, (Saunders), pink bollworm	T104
Potatoes	USA imports	*Graphognathus* spp., whitefringed beetles; *Ostrinia nubilalis* (Hübner), European corn borer; *Phthorimaea operculella* (Zeller), potato tuberworm	T104
Pumpkin	USA imports	Found on inspection	T101
Root crops Ginger	USA imports	Found on inspection	T104[a]

(Continues)

TABLE 5.1. (*Continued*)

Commodity	Destination	Pests of Regulatory Concern	Treatment Schedule
Sweet potatoes Yams	USA imports and Hawaii to mainland USA		T104[b]
HAWAIIAN FRUITS AND VEGETABLES FOR FRUIT FLIES			
Tomato	Hawaii to mainland USA	*Ceratitis capitata* (Wiedemann), Mediterranean fruit fly	T101,T105
GRAINS AND SEEDS			
Grain Ear corn Shelled corn	USA imports		T302[b]

[a]Treatment may be mandatory depending on country of origin.
[b]Treatment is mandatory regardless of the presence or absence of any pest.

TABLE 5.2. Commodities with pests of regulatory concern in the Export Certification Manual (APHIS 1986) that may require fumigation when exported from the USA to foreign countries

Country	Commodity	Pest of Regulatory Concern	Fumigant
Australia	*Abelmoschus esulentus*, Okra	*Pectinophora gossypiella* (Saunders)	CH_3Br
	Ananas comosus, pineapple	Miscellaneous pests	CH_3Br
	Citrullus lanatus, watermelon	*Bactrocera cucurbitae* (Coquillett), melon fly	CH_3Br
	Cocos nucifera, coconut	Miscellaneous pests	CH_3Br
	Cruciferae	*Pieris rapae* (L.) imported cabbageworm	CH_3Br
	Fragaria spp., strawberry	Miscellaneous pests	CH_3Br
	Nuts	Miscellaneous pests	CH_3Br, PH_3
Brazil	*Castanea* spp., chestnut	Miscellaneous pests	CH_3Br
	Zea mays, corn	Miscellaneous pests	CH_3Br, PH_3
British Colombia	*Chaenomeles* spp., flowering quince *Cydonia* spp., quince *Prunus armeniaca*, apricot *Prunus persica*, peach and nectarine *Prunus domestica*, plum	*Grapholita molesta* (Busck), oriental fruit moth	CH_3Br
	Gaylussaccia baccata *Gaylussaccia frondosa* *Gaylussaccia dumosa*	*Rhagoletis mendax* Curran, blueberry maggot	CH_3Br
	Sorghum bicolor *Sorghum sudanense* *Zea mays*, corn	*Ostrinia nubilalius* (Hübner), European corn borer	CH_3Br

(Continues)

TABLE 5.2. (*Continued*)

Country	Commodity	Pest of Regulatory Concern	Fumigant
Canada	*Vaccinium angustifolium*, lowbush blueberry *Vaccinium corymbosum*, highbush blueberry *Vaccinium oxycoccos*, small cranberry *Vaccinium viti-udaea* var. *minus*, cowberry or longonberry fruit	*Rhagoletis mendax* Curran, blueberry maggot	CH_3Br
Egypt	*Colocasia esculenta*, taro, eddo, dasheen	*Araecerus fasciculatus* (De Geer), coffee bean weevil	Unspecified
	Triticum, wheat *Zea mays*, corn, popcorn	Miscellaneous pests	Unspecified
India	*Zea mays*, corn	Miscellaneous pests	PH_3
Inter-African Group	*Prunus persica*, peach, nectarine	*Anastrepha ludens* (Loew), Mexican fruit fly; *Rhagoletis pomonella* (Walsh), apple maggot	Unspecified
Japan	*Carica papaya*, solo type papaya	Miscellaneous pests	Unspecified
	Citrus spp.	*Anastrepha suspensa* (Loew), Caribbean fruit fly	CH_3Br
	Juglans spp., walnut *Prunus persica*, nectarine *Prunus avium*, sweet cherry	*Cydia pomonella* (L.), codling moth	CH_3Br
	Punica spp., pomegranate	*Ceratitis capitata* (Wiedemann), Mediterranean fruit fly	CH_3Br
Mexico	*Zea mays*, corn	*Ostrinia nubilalis* (Hübner), European corn borer	CH_3Br

Country	Commodity	Pest	Treatment
New Zealand	*Allium cepa*, onion, shallot	*Delia antiqua* (Meigen), onion maggot	CH_3Br
	Ananas comosus, pineapple *Citrus* spp.	Coccoidea	Unspecified
	Diospyros spp., persimmon	Miscellaneous pests	Unspecified
	Fragaria spp., strawberry	Aphids; *Frankliniella occidentalis* (Pergande), western flower thrips; Miscellaneous pests; Mites	CH_3Br
	Malus spp., apple	*Rhagoletis pomonella* (Walsh), apple maggot	Unspecified
	Olea europaea, olive	Tephritid fruit flies	Unspecified
	Phoeniix dactylifera date *Prunus* spp., stone fruit except peach and nectarine *Punica granatum*, pomegranate	Tephritid fruit flies	CH_3Br
Republic of Korea	*Prunus* spp., cherries, 'Bing' 'Lambert,' 'Van'	*Cydia pomonella* (L.), codling moth	CH_3Br
Taiwan	*Anacardium occidentale*, cashew nut *Annona cherimola*, cherimoya *Annona reticulata*, custard apple *Annona squamosa*, sugar apple, sweetsop *Averrhoa carambola*, carambola *Citrus* spp.	*Anastrepha obliqua* (Macquart), West Indian fruit fly	Unspecified

(Continues)

TABLE 5.2. (Continued)

Country	Commodity	Pest of Regulatory Concern	Fumigant
	Dovyalis hebecarpa, Ceylon gooseberry, kitembilla *Eugenia* spp., grumichama, pitanga, Surinam cherry *Ficus carica,* fig *Inga laurina* *Lycopersicon lycopersicum,* tomato		
	Mammea americana, mamey, mammie apple, mammy apple *Mangifera indica,* mango *Prunus dulcis,* almond *Psidium guajava,* guava *Spondias purpurea* and *Spondias* spp.	*Anastrepha obliqua* (Macquart), West Indian fruit fly	
	Apium graveolens, celery *Beta vulgaris,* sugar beet *Brassica yoleracea,* cabbage *Brassica rapa,* turnip *Cucumis melo,* melon *Diospyros* spp., persimmon *Glycine max,* soybean *Ipomoea batatas,* sweet potato *Raphanus sativus,* radish	*Graphognathus leucoloma* (Boheman), whitefringed beetle	Unspecified
	Citrus spp. *Cydonia* spp., quince	*Anastrepha ludens,* (Loew), Mexican fruit fly	Unspecified

Host	Pest	
Malus spp., apple, crab apple Persea americana, avocado Pouteria sapota, sapota, zapote Prunus armeniaca, apricot Prunus domestica, plum Prunus persica, peach Pseudophoenix sargentii, Florida cherry palm Psidium quajava, guava Punica granatum, pomegranate Pyrus spp., pear	Conotrachelus nenuphar, (Herbst), plum curculio; Rhagoletis pomonella (Walsh), apple maggot	Unspecified
Crataegus spp., Hawthorne, haw, thorn apple; and Gaylussacia spp. huckleberry	Conotrachelus nenuphar (Herbst), plum curculio; Cydia pomonella (L.), codling moth	Unspecified Unspecified
Cydonia spp., quince	Cydia pomonella (L.), codling moth	Unspecified
Juglans spp., walnut	Cydia pomonella (L.), codling moth	Unspecified
Malus spp., apple, crab apple	Conotrachelus nenuphar (Herbst), plum curculio; Cydia pomonella (L.), codling moth; Rhagoletis pomonella (Walsh), apple maggot	Unspecified
Prunus armeniaca, apricot and P. domestica, plum	Anarsia lineatella Zeller, peach twig borer; Conotrachelus nenuphar (Herbst), plum curculio; Cydia pomonella (L.), codling moth; Rhagoletis pomonella (Walsh), apple maggot	Unspecified
Prunus avium, sweet cherry and P. cerasus, sour cherry	Anarsia lineatella Zeller, peach twig borer; Conotrachelus nenuphar; (Herbst), plum curculio; Cydia pomonella (L.), codling moth	Unspecified Unspecified

(Continues)

TABLE 5.2. (*Continued*)

Country	Commodity	Pest of Regulatory Concern	Fumigant
	Prunus persica, peach	*Anarsia lineatella* Zeller, peach twig borer; *Conotrachelus nenuphar* (Herbst), plum curculio; *Cydia pomonella* (L.), coding moth; *Rhagoletis pomonella* (Walsh), apple maggot	Unspecified
	Prunus persica var. *nucipersica*, nectarine	*Anarsia lineatella* Zeller, peach twig borer; *Conotrachelus nenuphar* (Herbst), plum curculio	Unspecified
	Pyrus spp., pear	*Conotrachelus nenuphar*, (Herbst), plum curculio; *Cydia pomonella* (L.), coding moth; *Rhagoletis pomonella* (Walsh), apple maggot;	Unspecified
	Vaccinium spp., blueberry	*Conotrachelus nenuphar* (Herbst), plum curculio; *Rhagoletis pomonella* (Walsh), apple maggot	Unspecified
Venezuela	*Cucumis melo*, cantaloupe *Cucumis sativus*, cucumber *Cydonia oblonga*, quince *Lycopersicom esculentum*, tomato *Malus sylvestris*, apple *Prunus americana*, plum *Prunus armeniaca*, apricot *Prunus avium*, sweet cherry *Prunus cerasus*, sour cherry *Prunus persica*, peach *Pyrus communis*, pear *Solanum melongena*, eggplant *Vitis* spp., grape	Miscellaneous pests	Unspecified

TABLE 5.3. Exported commodities and pests of regulatory concern for which methyl bromide fumigation has been developed as a quarantine treatment but has not yet been approved

Commodity	Destination	Pests of Regulatory Concern	Reference
Carambola	USA interstate and exports	*Anastrepha suspensa* (Loew), Carribbean fruit fly	Hallman & King (1992)
Citrus	USA to Japan	*Asynonychus godmani* Crotch, Fuller rose beetle	Soderstrom et al. (1991)
Cucumber	Hawaii to mainland USA	*Bactrocera cucurbitae* (Coquillett), melon fly *Bactrocera dorsalis* (Hendel), oriental fruit fly	Armstrong & Garcia (1985)
Stone fruits	USA to New Zealand	*Rhagoletis completa* Cresson, walnut husk fly	Yokoyama et al. (1992)
Strawberry	Hawaii exports	*Ceratitis capitata* (Wiedemann), Mediterranean fruit fly	Armstrong et al. (1984)

TABLE 5.4. Exported and imported commodities with pests of regulatory concern for which methyl bromide fumigation has been developed as a quarantine treatment and residues have been reported

Commodity	Destination	Pests of Regulatory Concern	Reference
Apple	USA to Japan	*Cydia pomonella* (L.), codling moth	Sell & Moffitt (1990)
Asparagus Avocado Lychee Orange Papaya Pepper Pineapple Tomato	Hawaii to mainland USA	*Ceratitis capitata* (Wiedemann), Mediterranean fruit fly; *Bactrocera dorsalis* (Hendel), oriental fruit fly; *Bactrocera cucurbitae* (Coquillett), melon fly	Seo et al. (1970)
Cherry Nectarine Peach Pear Plum	Hawaii to mainland USA	*Ceratitis capitata* (Wiedemann), Mediterranean fruit fly	Tebbets et al. (1983)
Grapefruit	USA interstate and exports	*Anastrepha suspensa* (Loew), Caribbean fruit fly	King et al. (1981)
Mango	Mexico to USA	*Anastrepha ludens* (Loew), Mexican fruit fly	Stein & Wolfenbarger (1989)
Potato	USA to United Kingdom	*Phthorimaea operculella* (Zeller), potato tuberworm	Gostick et al. (1971)

References

Anonymous. 1984. Environmental Protection Agency. Rules and regulations. Revocation of tolerance ethylene dibromide. Federal Register 49: 22082-22085.

_____. 1986. Export certification manual, Vols. 1-6. U. S. Government Printing Office. Washington, D.C.

_____. 1992a. *Montreal Protocol Assessment Supplement. Methyl Bromide: Its Atmospheric Science, Technology and Economics. Synthesis Report of the Methyl Bromide Interim Scientific Assessment and Methyl Bromide Interim Technology and Economic Assessment.* United Nations Environment Programme. U.S. Government Printing Office. Washington, D.C.

_____. 1992b. Animal and Plant Health Inspection Service. Plant protection and quarantine treatment manual. U.S. Government Printing Office. Washington, D.C.

Armstrong, J. W. & D. L. Garcia. 1985. Methyl bromide quarantine fumigations for Hawaii-grown cucumbers infested with melon fly and oriental fruit fly (Diptera: Tephritidae). J. Econ. Entomol. 78: 1308-1310.

Armstrong, J. W., E. L. Schneider, D. L. Garcia & H. M. Couey. 1984. Methyl bromide quarantine fumigation for strawberries infested with Mediterranean fruit fly (Diptera: Tephritidae). J. Econ. Entomol. 77: 680-682.

Bond, E. J. 1984. *Manual of Fumigation for Insect Control.* 432 pp. Rome, Italy: Food and Agr. Organization of the United Nations.

Fiskaali, D. A. 1989. *The State of California Commodity Treatment Manual, Vol. I, Treatments.* Sacramento, California: State of California Department of Food and Agr.

Gaunce, A. P., H. F. Madsen, R. D. McMullen & J. W. Hall. 1980. Dosage response of the stages of codling moth, *Laspeyresia pomonella* (Lepidoptera: Olethreutidae), to fumigation with methyl bromide. Can. Entomologist 112: 1033-1038.

Gostick, K. G., S. G. Heuser, G. Goodship & D. F. Powell. 1971. Bromide residues in potatoes fumigated with methyl bromide. Potato Research 14: 312-315.

Hallman, G. J. & J. R. King. 1992. Methyl bromide fumigation quarantine treatment for carambolas infested with Caribbean fruit fly (Diptera: Tephritidae). J. Econ. Entomol. 85: 1231-1234.

Hartsell, P. L., P. V. Vail, J. S. Tebbets & H. D. Nelson. 1991a. Methyl bromide quarantine treatment for codling moth (Lepidoptera: Tortricidae) in unshelled walnuts. J. Econ. Entomol. 84: 1289-1293.

Hartsell, P. L., J. C. Tebbets & P. V. Vail. 1991b. Methyl bromide residues and desorption rates from unshelled walnuts fumigated with a quarantine treatment for codling moth (Lepidoptera: Tortricidae). J. Econ. Entomol. 84: 1294-1297.

Harvey, J. M., C. M. Harris & P. L. Hartsell. 1989. Tolerances of California nectarine cultivars to methyl bromide quarantine treatments. J. American Soc. Hortic. Sci. 114: 626-629.

King, J. R., C. A. Benschoter & A. K. Burditt, Jr. 1981. Residues of methyl bromide in fumigated grapefruit determined by a rapid headspace assay. J. Agric. Food Chemistry 29: 1003-1005.

LeOra Software Inc. 1987. POLO-PC. LeOra Software, Inc., 1119 Shattuck Ave., Berkeley, California.

Robertson, J. L. & H. K. Priesler. 1992. *Pesticide Bioassays with Arthropods*. Boca Raton, Florida: CRC Press

SAS Institute, Inc. 1988. SAS/Stat User's guide, release 6.03 edition. SAS Institute, Cary, North Carolina.

Sell, C. R. & H. R. Moffitt. 1990. Non-destructive method for estimating methyl bromide residues in apples during aeration following fumigation. Pesticide Sci. 29: 19-27.

Seo, S. T., J. W. Balock, A. K. Burditt, Jr., & K. Ohinata. 1970. Residues of ethylene dibromide, methyl bromide, and ethylene chlorobromide resulting from fumigation of fruits and vegetables infested with fruit flies. J. Econ. Entomol. 63: 1093-1097.

Soderstrom, E. L., D. G. Brandl, P. L. Hartsell & B. Mackey. 1991. Fumigants as treatments for harvested citrus fruits infested with *Asynonychus godmani* (Coleoptera: Curculionidae). J. Econ. Entomol. 84: 936-941.

Stein, E. R. & D. A. Wolfenbarger. 1989. Methyl bromide residues in fumigated mangos. J. Agric. Food Chemistry 37: 1509-1513.

Tebbets, J. S., P. L. Hartsell, H. D. Nelson & J. C. Tebbets. 1983. Methyl bromide fumigation of tree fruits for control of the Mediterranean fruit fly: concentrations, sorption, and residues. J. Agric. Food Chemistry 31: 247-249.

Ware, G. W. 1978. *The Pesticide Book*. San Francisco, California: W. H. Freeman.

Yokoyama, V. Y., G. T. Miller & P. L. Hartsell. 1987a. Methyl bromide fumigation to control the oriental fruit moth (Lepidoptera: Tortricidae) in nectarines. J. Econ. Entomol. 80: 1226-1228.

_____. 1987b. Methyl bromide fumigation for quarantine control of codling moth (Lepidoptera: Tortricidae) on nectarines. J. Econ. Entomol. 80: 840-842.

_____. 1988. Rearing, large-scale tests, and egg response to confirm efficacy of a methyl bromide quarantine treatment for codling moth (Lepidoptera: Tortricidae) on exported nectarines. J. Econ. Entomol.

81: 1437-1442.

_____. 1990a. Evaluation of a methyl bromide quarantine treatment to control codling moth (Lepidoptera: Tortricidae) on nectarine cultivars proposed for export to Japan. J. Econ. Entomol. 83: 466-471.

_____. 1990b. A methyl bromide quarantine treatment to control codling moth (Lepidoptera: Tortricidae) on nectarines packed in shipping containers for export to Japan and effect on fruit attributes. J. Econ. Entomol. 83: 2335-2339.

_____. 1992. Pest-free period and methyl bromide fumigation for control of walnut husk fly (Diptera: Tephritidae) in stone fruits exported to New Zealand. J. Econ. Entomol. 85: 150-156.

_____. 1993. Methyl bromide efficacy and residues in large-scale quarantine tests to control codling moth (Lepidoptera: Tortricidae) on nectarines in field bins and shipping containers for export to Japan. J. Econ. Entomol. 87: (in press).

6

Pesticide Quarantine Treatments

Neil W. Heather

The scope for pesticides as quarantine treatments against pests of food plants is broad, ranging from field application as pest suppression and eradication programs to postharvest disinfestation treatments. However, pesticides which can be used are limited by regulatory considerations that vary among countries. The main postharvest uses of pesticides for quarantine purposes have been in Australia as disinfestation treatments for Tephritidae flies. Treatments can be dips or sprays which may be applied to fruits and vegetables as they are processed. The purpose of the chapter is to discuss the use of pesticides in quarantine disinfestations of pests from fruits and vegetables.

Safety Standards

Pesticide usage must meet relevant health and safety standards with respect to application and residues. Safety of operators and the public in the application of pesticides is subject to regulatory control by health and agricultural authorities according to the state or country. Establishing legal residue limits in food requires the coordination of toxicological studies which are used to set an acceptable daily intake (ADI) by humans for the pesticide and residue studies to provide data to set maximum residue limits (MRLs).

The United Nations Food and Agricultural Organization, World Health Organization (FAO/WHO), Codex Alimentarius Commission, established a mechanism for governments to agree on MRLs, thus facilitating world trade in food commodities. Countries have either accepted FAO/WHO estimates of ADIs and MRLs or established their own. Use of a pesticide is permitted only if residues from that use are compatible with the ADI (Hamilton 1988).

Criteria for Effectiveness

Postharvest disinfestation treatments must achieve efficacy standards which relate to recognized levels of quarantine security, be practical to handle and apply, and not reduce fruit quality. A security level required by United States Department of Agriculture is probit 9 or 99.9968% mortality (Chew & Ouye 1985). Probit 9 gives a maximum pest limit (MPL) of 32 survivors from a million of the most tolerant treated stage in or on the fruit. Other countries have different standards. For example, New Zealand has an MPL for fruit flies of five in a million fruits based on the data of Baker et al. (1990). This standard is achieved partly through disinfestation treatments, which may be physical or chemical. In tomatoes, reduction to five fruit flies in a million treated fruits is accomplished through a combination of pest management procedures in the production system subject to an agreed protocol which includes a 1 min postharvest dip in dimethoate. This system requires registration of growers and defines field inspection procedures and control sprays to be undertaken while crops are being grown. It also details packing house inspections to be undertaken by the producer, postharvest treatments, and formal sampling rates by Australian and New Zealand quarantine service inspectors, together with rejection levels for other pests.

Standards of required efficacy vary among countries and are conventionally demonstrated by confirmatory trials against, for example, 93,616 insects for probit 9 or 29,956 for probit 8.7, at the 95% confidence level (Couey & Chew 1986). Some countries may require the treatment to be shown to have no survivors in three trials each on 10,000 or four trials each on 7,500 insects of the most tolerant stage (Australian Plant Health Inspection Service, personal communication).

Insecticide dips and packing line flood sprays have achieved a 99.99% level of efficacy against fruit flies in thin skinned fruits such as tomatoes (Swaine et al. 1984a, Heather et al. 1987). On thicker skinned fruit such as mangoes, the efficacy is lower (Swaine et al. 1984b). There are no treatments with insecticide dips or sprays with a proven efficacy level of 99.9968% in commercial use against fruit flies. The results of Saul

& Seifert (1990) indicate this capability for the insect growth regulator (IGR), methoprene.

Pesticides

Organophosphorus

The use of dimethoate (De Pietri-Tonelli & Barontini 1957) and fenthion (Unterstenhofer 1960), insecticides with significant systemic action against the eggs and larvae of fruit flies in fruit, created a new approach to field control of these pests (May 1962). Both insecticides are used in Australia as postharvest disinfestation treatments for a range of fruits. Their postharvest efficacy ranges from >99.5 to >99.99% at the 95% confidence level. A 1 min, 400 mg per liter dimethoate dip disinfestation treatment against Queensland fruit fly, *Bactrocera tryoni* (Froggatt), was first accepted for a range of fruits and vegetables by the State of Victoria (Anonymous 1982).

Braithwaite (1963) reported postharvest application trials in which bananas were dipped for 1 min in 500 mg per liter of dimethoate or fenthion in water as disinfestation treatments against Queensland fruit fly. Resulting residues for dimethoate at all times were <2 mg per liter, the Australian MRL (National Health and Medical Research Council 1988). No Australian MRL exists for fenthion on bananas. A commercial treatment unit applying a spray of 300 mg per liter of dimethoate was used for a number of years in Queensland. Saunders & Elder (1966) and Smith (1977) reported similar trials against banana fruit fly, *Bactrocera musae* (Tryon). In Taiwan, Lee (1968) reported on the efficacy of three organophosphorus insecticides against oriental fruit fly, *Bactrocera dorsalis* (Hendel), with best results from a 1 min dip in fenitrothion. Fenitrothion is not claimed to have a markedly penetrating mode of action.

Swaine et al. (1984a, b) tested dimethoate on tomatoes and mangoes. For tomatoes a dip for 1 min in 425 mg per liter of dimethoate caused 99.997 and 100% mortality in >30,000 eggs 24 h old and >30,000 B. *tryoni* larvae 5 d old, respectively. A 3 min dip caused 100% mortality on both eggs and larvae with residues that averaged 0.58 mg per kg on the day of treatment. For mangoes, a 3 min dip in 500 mg per liter of dimethoate caused 99.98 and 99.97% mortality in >30,000 eggs 24 h old and >30,000 larvae 5 d old, respectively. When the insecticide was mixed with 55°C water and benlate to control anthracnose disease, insect mortality was slightly lower compared with dips at ambient temperature. Resultant residues of dimethoate were always below the Australian MRLs of 1 mg per kg for tomatoes and 2 mg per kg for mangoes. Adverse tastes were not detected in any dipped fruit of either tomatoes or mangoes, and fruit quality was not affected. Swaine et al. (1984a) observed that in tomatoes third instars were less susceptible than eggs, and individuals occasionally pupated. Adults did not normally emerge from the puparia. Quarantine security was unaffected.

Heather et al. (1987) reported efficacy levels against *B. tryoni* >99.99% for high volume recirculatory flood treatments of 400 mg per liter of dimethoate or fenthion on tomatoes. Wetting times were equivalent to a 1 min dip. Neither treatment resulted in residues in excess of the Australian MRLs for fenthion (2 mg per kg) or dimethoate, nor were there any adverse effects on taste. Subsequent trials with these insecticides on rockmelons and zucchinis against cucumber fly, *Bactrocera cucumis* (French), gave similar results, again with no adverse taste effects (Heather et al. 1992).

Dip and flood spray treatments are alternatives to fumigation, especially for fruits or vegetables prone to phytotoxicity. The dimethoate dip for tomatoes was developed primarily because they were susceptible to damage when fumigated with ethylene dibromide. Insecticide treatments also provide residual protection against subsequent infestation.

Dip or recirculatory flood spray systems could increase the risk of spreading inoculum of postharvest rots, although this has not proved to be a problem in practice. Three other risks to the reliability of insecticide dips or flood sprays in closed systems are perceived. The first is "stripping" of the insecticide, which occurs when the active ingredient is selectively removed from the dip formulation by adherence to the fruit (a characteristic of some fungicides). Neither the dimethoate dip life study by Noble (1983) nor measurements on experimental spray vats before and after fruit treatment (Heather, unpublished data) showed loss of active ingredient. The second risk to insecticide concentrations can occur where fruits are washed then dipped or sprayed while still wet. Over time, water is added to the system which dilutes the dip. The third risk, chemical decay of the active ingredient, can be predicted and compensated for by adding more insecticide periodically (Noble 1983).

Field applications of dimethoate or fenthion may be appropriate to meet quarantine security. An example is quarantine requirements for interstate trade in Australia between Queensland, where fruit flies are endemic, and Victoria, which is at the southernmost limit of distribution and virtually free of fruit flies (Bateman 1967). Here, quarantine security is met in some circumstances with a grower's declaration that a dimethoate or fenthion spray was applied before harvest. This technique has proved effective for those crops in which good spray coverage of each fruit is possible. It can fail where fruit fly populations are high and spray coverage is insufficient.

Ethylene Dibromide

Ethylene dibromide (EDB) was used in many countries as a postharvest fumigation treatment against fruit flies. The chemical was incorporated in a hot water dip (Burditt et al. 1963). Probit 9 mortality against oriental fruit fly, melon fly, *Bactrocera cucurbitae* (Coquillett), and Mediterranean fruit fly, *Ceratitis capitata* (Wiedemann), was achieved by dipping papayas for 20 min in 46°C water containing 108.5 mg per liter of EDB. Residues were 0.1 mg per kg after 3 d. The hot water dip combined with EDB was never used commercially, and postharvest uses of EDB were banned in the United States (U.S.) in 1984 (Anonymous 1984). Effectiveness of the treatment of Burditt et al. (1963) may have been due in part to the water temperature, although Lee (1968) reported from Taiwan that a similar treatment in water at ambient temperature was effective against the most tolerant species, *B. dorsalis*, in bananas. Subsequently, an EDB dip of 1 - 1.3 mg per liter for 3 min and 20 sec against *B. dorsalis* and *B. cucurbitae* was accepted for export from Taiwan to Japan from 1969 to 1978 (Tseng et al., in press).

Organochlorine

Endosulfan is approved in Australia as a postharvest dip to disinfest pineapples of dried fruit beetles, *Carpophilus* species (Beavis et al. 1991). A 1 min dip in 560 mg per liter of endosulfan in water enables fruit to meet an export inspection tolerance of no infested fruit per 600 sample for an MPL of 0.5% at the 95% confidence level (Couey & Chew 1986). Residues measured on the day of treatment were 1.9 mg per kg and, 7 d later, 1.4 mg per kg (Queensland Department of Primary Industries, unpublished data). The current Australian MRL for endosulfan on fruit is 2 mg per kg.

Dicofol is a miticide that has potential for use as a postharvest or immediately preharvest treatment against mites that attack fruits and vegetables. The Australian MRL is 5 mg per kg on fruit. Cosmopolitan mites are pests for which probit 9 mortality is rarely required from disinfestation treatments. However when zero or low tolerance levels apply at inspection, such as the maximum of 20 infested fruit in a sample of 600 (an MPL of 5%) used in Australia and New Zealand protocols, preharvest or postharvest population suppression measures are frequently required.

Insect Growth Regulators

Disinfestation of papayas against *C. capitata*, *B. dorsalis*, and *B.*

cucurbitae with the IGR methoprene was reported by Saul & Seifert (1990). Their study identified late third instars as the most susceptible stage and eggs the most resistant. Methoprene (20%) formulated in wax and applied postharvest to fruit infested with eggs and larvae achieved probit 9 mortality on the basis of failure to develop to adults. Therefore, problems could arise if larvae were detected during subsequent inspections. This situation is similar to irradiation and, to a lesser extent, organophosphorus dip treatments, where live larvae that were properly treated and eventually die might be found. Strict regulation of pest management programs during production and packing can overcome the problem by preventing detectable infestation of export produce. Use of IGRs is a new approach to quarantine disinfestation both in the pesticide type used and the method of application, i.e., the chemical is mixed with the wax that is applied during packing to protect fruit and to enhance their appearance.

Application

Commercial Application

In commercial operations insecticide dipping is typically done by immersing fruit held in a wire basket in a tank of dilute insecticide (Fig. 6.1). For quantities of about one cubic meter, a fork lift is used. Smaller quantities can be safely dipped manually. Export fruit should be washed, dried, culled, and graded prior to dipping because the disinfestation dip must be the last treatment applied. No chemical which could cause breakdown of the insecticide (e.g., sodium hypochlorite in an organophosphorus mix) should be allowed to contaminate the dip. Accurate timing of immersion ensures that efficacy is not impaired or the MRL exceeded. The same applies to flood spray application (Fig. 6.2). Both methods are most widely used in Queensland for tomatoes exported to fruit fly free Australian states or to New Zealand. For the most commonly used insecticide, dimethoate, the maximum usage of a dip should not exceed one month. During this time, concentrate is to be added every 7 d to compensate for decay of the active ingredient in accordance with its half life of 148 d at pH 6 and 25°C (Noble 1983).

Combination with Other Treatments

Combining insecticides with other treatments to disinfest fruits and vegetables has not been fully exploited. Hot water dips for disease

FIGURE 6.1.   A commercial quarantine dipping operation with dimethoate at 400 mg per liter. The product (tomatoes) is fully immersed for 1 min.

FIGURE 6.2. A commercial grading machine module set up to apply insecticide. The first spray bar is typically used as a water wash, the second to apply fungicide or insecticide at low volume, and the third for high volume flood application of insecticide. Excess from the low volume spray is usually not reused but the flood system is recirculated. Fruit may be multiply treated in a single pass but it is also possible to use multiple passes with different treatments applied each time. (Reproduced with permission of the Editor, *Australian Journal of Experimental Agriculture* .)

control combined with low postharvest storage temperatures cause mortality to fruit flies (Armstrong & Couey 1989). The addition of an insecticide to the hot water dip may provide quarantine security, although only thermostable insecticides can be used in this way. The relatively high postharvest storage temperatures required by many tropical fruits do not kill fruit fly eggs and larvae as quickly as do colder postharvest temperatures tolerated by many temperate fruits. For example, holding temperatures are 10°C for papaya (Armstrong & Couey 1989) and 13°C for tomatoes (McGlasson et al. 1982) compared with 3°C for citrus (Armstrong & Couey 1989). However, insecticide added to hot water for disease control or applied separately could be used to increase mortality at postharvest storage temperatures to meet many disinfestation security requirements (Heather et al. 1987).

Other Applications

Insecticides are commonly used for quarantine purposes against pests other than fruit flies. Dips can be used to disinfest planting stock, where, because residues are not a problem for plant material that is not consumed, insecticides of higher toxicities can be used. In Australia, diazinon is used to disinfest citrus nursery stock of leaf miner, *Phyllocnistis citrella* Stainton, and on various plant materials against cattle tick, *Boophilus microplus* (Canestrini) (Anonymous 1982). Field applications of 3000 liters per hectare of a 0.5 g per liter spray of chlorpyrifos or diazinon are used to disinfest pineapples of pineapple mealybug, *Dysmicoccus brevipes* (Cockerell), (Beavis et al. 1991), thus enabling the New Zealand quarantine MPL of 0.5% to be met. In the U.S. azinphos-methyl or carbaryl may be used as foliar sprays to suppress populations of Fuller rose beetle, *Asynonychus godmani* Crotch. This is a relatively innocuous pest of citrus that is subject to quarantine action (fumigation or rejection of the shipment) when eggs are found under the calyx of fruit exported to Japan (Morse et al. 1988). Foliar applications, combined with cultural practices such as skirt pruning, provide quarantine security for citrus exported to Japan without sacrificing integrated pest management programs which minimize pesticide residues on the fruit.

Conclusions

Where the criteria of safety, efficacy, and practicality can be met, pesticides offer realistic treatments for postharvest quarantine disinfestation. The major disadvantages are residues and toxicity risks to operators. The advantages of insecticides include economical costs, logistical flexibility, simplicity of application, residual protection, and ease of supervision. They are particularly appropriate where the efficacy required is less than probit 9 and, at lower levels of security, may achieve quarantine security when applied preharvest. They can also be an important component of combination systems which, in total, meet efficacy requirements of probit 9 or higher equivalent MPLs.

There is a role for pesticides in both postharvest disinfestation and pest suppression in the field as adjuncts to quarantine systems which include pest management requirements to meet established MRLs. Both usages are subject to the existence of achievable MRLs. Where safety margins are adequate, some adjustment of MRLs may be justified given their relationship to good agricultural practice rather than to public health risk alone. If MRLs exist for field use of the same pesticides postharvest, no different residues are then being introduced to the product. Residues are more accurately controlled in postharvest application through precise concentration and timing of the dip or spray compared with field application. In Australia the National Residue Survey (Anonymous 1989) recorded a majority of samples with no detectable residues, a situation expected to be true for most agricultural and horticultural production systems. However, few edible plant products marketed today are always entirely free of pesticide residues, an accepted trade off for improved quality.

Acknowledgments

Helpful comment on the manuscript by J. W. Armstrong, United States Department of Agriculture, Agricultural Research Service, Hilo, Hawaii, A. K. Burditt, Jr., Ocala, Florida, formerly with the United States Department of Agriculture, Agricultural Research Service, and D. Hamilton, Department of Primary Industries, Indooroopilly, Queensland, Australia, was greatly appreciated.

References

Anonymous. 1982. Statutory rules, vegetation and vine diseases act (State of Victoria). Government Printer, Melbourne.

_____. 1984. Environmental Protection Agency. Rules and regulations. Revocation of tolerance for ethylene dibromide. Federal Register 49 (103): 22082-5.

_____. 1989. Report on national residue survey: 1987 & 1988 results. Bureau of Rural Resources. Australian Government Publishing

Service, Canberra.

Armstrong, J. W. & H. M. Couey. 1989. "Fumigation, Heat and Cold," in A. S. Robinson & G. Hooper, eds., *World Crop Pests, Vol. 3B, Fruit Flies. Their Biology, Natural Enemies and Control.* Pp. 411-424. Amsterdam: Elsevier.

Baker, R. T., J. M. Cowley, D. S. Harte & E. R. Frampton. 1990. Development of a maximum pest limit for fruit flies (Diptera: Tephritidae) in produce imported into New Zealand. J. Econ. Entomol. 83: 13-17.

Bateman, M. A. 1967. Adaptations to temperature in geographic races of the Queensland fruit fly, *Dacus (Strumeta) tryoni.* Australian J. Zoology 15: 1141-61.

Beavis, C., P. Simpson, J. Syme & C. Ryan. 1991. Chemicals for the protection of fruit and nut crops. Department of Primary Industries, Queensland Government, Brisbane.

Braithwaite, R. M. 1963. The sterilisation of banana fruit against Queensland fruit fly (*Strumeta tryoni*). Australian J. Experimental Agr. & Animal Husbandry 3: 98-100.

Burditt, A. K., Jr., J. W. Balock, F. G. Hinman & S. T. Seo. 1963. Ethylene dibromide water dips for destroying fruit fly infestations of quarantine significance in papayas. J. Econ. Entomol. 56: 289-292.

Chew, V. & M. T. Ouye. 1985. "Statistical Basis for Quarantine Treatment Schedule and Security," in J.H. Moy, ed., *Radiation Disinfestation of Food and Agricultural Products, Proceedings of an International Conference, Honolulu (1983).* Pp. 70-74. Honolulu, Hawaii: Hawaii Institute of Tropical Agr. and Human Resources, University of Hawaii at Manoa.

Couey, H. M. & V. Chew. 1986. Confidence limits and sample size in quarantine research. J. Econ. Entomol. 79: 887-890.

De Pietri-Tonelli, P. & A. Barontini. 1957. Activity of N-monomethyla-mide of O, O-dimethyl dithiophosphorylactic acid (L395), of parathion and their mixture with copper sulphate against *Dacus oleae.* Olivicoltura 12: 6 (from Chemical Abstracts 1957, 17060i).

Hamilton, D. 1988. Setting MRL values for pesticides in foods. Queensland Agric. J. 114: 225-228.

Heather, N. W., P. A. Hargreaves, R. J. Corcoran & K. J. Melksham. 1987. Dimethoate and fenthion as packing line treatments for tomatoes against *Dacus tryoni.* Australian J. Experimental Agr. 27: 465-469.

Heather, N. W., D. E. Walpole, R. J. Corcoran, P. A. Hargreaves & R. A. Jordan. 1992. Postharvest quarantine disinfestation of zucchinis and rockmelons against *Bactrocera cucumis* (French) using insecticide dips of fenthion or dimethoate. Australian J. Experimental Agr. 32: 241-244.

Lee, S. 1968. Effectiveness of various insecticides for destroying oriental fruit fly, *Dacus dorsalis* Hendel, infestations in green mature banana. J. Taiwan Agric. Res. 17: 17-25.

May, A.W.S. 1962. The fruit fly problem in eastern Australia. J. Entomological Soc. Queensland 1: 1-4.

McGlasson, W. B., R. L. McBride, D. J. McGrath, E. F. Smith & D. J. Best. 1982. Influence of harvest maturity and ripening temperature on acceptability of North Queensland fresh market tomatoes. Food Technology Australia 34: 291-293.

Morse, J. G., D. A. De Mason, M. S. Arpaia, P. A. Phillips, P. B. Goodell, A. A. Urena, P. B. Haney & D. J. Smith. 1988. Options in controlling the Fuller rose beetle. Citrograph 73: 135-140.

National Health and Medical Research Council. 1988. MRL standard. Standard for maximum residue limits of pesticides, agricultural chemicals, feed additives, veterinary medicines and noxious substances in food. Australian Government Publishing Service, Canberra.

Noble, A. 1983. Fruit and vegetables - stability of dimethoate in dips. Refnote R6 Agdex/681. Queensland Department of Primary Industries, Brisbane.

Saul, S. H. & J. Seifert. 1990. Methoprene on papayas - Persistence and toxicity to different development stages of fruit flies (Diptera: Tephritidae). J. Econ. Entomol. 83: 901-904.

Saunders, G. W. & R. J. Elder. 1966. Sterilisation of banana fruit infested with banana fruit fly. Queensland J. Agric. Sci. 23: 81-85.

Smith, E.S.C. 1977. Studies on the biology and commodity control of the banana fruit fly, *Dacus musae* (Tryon) in Papua, New Guinea. Papua, New Guinea Agric. J. 28: 47-56.

Swaine, G., P. A. Hargreaves, D. E. Jackson & R. J. Corcoran. 1984a. Dimethoate dipping of tomatoes against Queensland fruit fly *Dacus tryoni* (Froggatt). Australian J. Experimental Agr. & Animal Husbandry 24: 447-449.

Swaine, G., K. J. Melksham & R. J. Corcoran. 1984b. Dimethoate dipping of Kensington mango against Queensland fruit fly. Australian J. Experimental Agr. & Animal Husbandry 14: 620-623.

Tseng, Y. H., L. S. Kud & J. Z. Ho. (In press). "Development of Quarantine Sterilization Techniques on Fruits in Taiwan." *Proceedings of Asian Productivity Organisation Study Meeting on Plant Quarantine, Taichung and Taipei, March, 1992.* Asian Productivity Organisation, Tokyo.

Unterstenhofer, G. 1960. Lebaycid, a new insecticide of low toxicity. Hofchen-Briefe 13: 44-52.

7

Irradiation

Arthur K. Burditt, Jr.

Irradiation is the process of treating commodities with ionizing energy and includes radio waves, radar, microwaves, infrared, visible rays, ultraviolet rays, X rays, and gamma rays. In the United States (U.S.), the Food and Drug Administration (FDA) approved the use of ionizing radiation, including gamma radiation from cobalt 60 and cesium 137, high energy electrons derived from an electron beam accelerator of ≤10 MeV, and X rays from a source with a beam energy of ≤5 MeV for food irradiation at a dose of ≤1,000 Gy (Anonymous 1984a). The energies from cobalt 60 gamma rays are 1.17 and 1.33 MeV. Energy from cesium 137 is 0.662 MeV. The U.S. Department of Agriculture, Animal and Plant Health Inspection Service, Plant Protection and Quarantine (USDA-APHIS-PPQ), authorized irradiation as a quarantine treatment for papayas intended for movement from Hawaii to the continental U.S., Guam, Puerto Rico, and the U.S. Virgin Islands to prevent introduction of oriental fruit fly, *Bactrocera dorsalis* (Hendel), Mediterranean fruit fly, *Ceratitis capitata* (Wiedemann), and melon fly, *Bactrocera cucurbitae* (Coquillett) (Anonymous 1989a). Gamma rays and X rays are efficient ionizing energies for quarantine purposes because they easily penetrate the commodity. Ionizing radiation used to treat commodities of quarantine importance is safe and will not induce radioactivity. No scientific evidence exists that human toxicological problems occur following treatment of foods at doses ≤10,000 Gy (Anonymous 1981).

Use of irradiation as a quarantine treatment requires precise control and verification of dose by using an automated system that includes product identification information and a dosimeter sufficiently accurate to measure the minimum dose of required irradiation (Prusik & Wallace

1985). The minimum dose must be applied to the entire batch of a commodity being treated.

History

Koidsumi (1930) suggested that X rays could be used as a quarantine treatment over 60 years ago. Balock et al. (1956, 1966) proposed that gamma rays from cobalt 60 could be used as a commodity treatment for Hawaiian fruit flies. Macfarlane (1966), Shipp & Osborn (1968), and Eric et al. (1970) proposed irradiation as a treatment for Queensland fruit fly, *Bactrocera tryoni* (Froggatt), in Australia.

During the last 20 years the Food and Agriculture Organization (FAO) and International Atomic Energy Agency (IAEA) have taken the lead in coordinating research on use of irradiation as a quarantine treatment. In 1970 the FAO and IAEA convened a panel on the use of irradiation to solve quarantine problems in the international fruit trade, "to collate existing knowledge of radiation technique and to define future activities for overcoming existing quarantine barriers in the international fruit trade by this method" (Anonymous 1971). The panel identified the potential for irradiation as a quarantine disinfestation treatment for fruit, primarily against fruit flies and mango weevil, *Cryptorhynchus mangiferae* (F.), identified the doses required for quarantine treatment of infested fruit, discussed criteria required for quarantine security, and considered both the effects of irradiation on fruits and methods to implement the technique for disinfestation by irradiation. The panel urged that the FAO and IAEA promote wholesomeness tests, foster pilot-plant studies to determine the feasibility of disinfestation by ionizing radiation, and encourage development of uniform pest quarantine requirements.

A coordinated research program on insect disinfestation of food and agricultural products by irradiation initiated by the FAO and IAEA in 1981 recommended that research be conducted on the development of general quarantine doses for particular pests regardless of host commodity. Also, the program recommended that a code of practice be prepared on the use of irradiation for inclusion in the FAO Quarantine Treatment Manual, complete studies to define the relationships between commodities, develop data on phytotoxicity, and organize an international training course for quarantine inspectors (Anonymous 1984c, 1985). A consultants group on the use of irradiation as a quarantine treatment of agricultural commodities recommended that the FAO and IAEA establish and implement a research coordination program (RCP) on the use of irradiation as an alternative quarantine treatment, use inability to produce viable offspring as the criterion for quarantine security, evaluate phytotoxic reactions of commodities that have been treated by irradiation, develop a manual for quarantine treatment by the radiation process, and encourage member countries to adopt the use of irradiation as a quarantine treatment (Anonymous 1984b). The RCP on the use of irradiation as a quarantine treatment of food and agricultural commodities was implemented (Anonymous 1986b, 1989b, 1991b). The RCP recommended that, (1) a bibliography of key papers be compiled, (2) a test be developed to verify that a pest has been irradiated and is unable to reproduce, (3) studies be initiated to determine effectiveness of quarantine protocol profiles for pests and tolerance of host plants, (4) training programs be developed for research, plant protection, and quarantine regulatory personnel, and (5) research be conducted on the use of electron beams and X rays.

The International Consultative Group on Food Irradiation (ICGFI) was established in 1984 by the FAO, IAEA, and World Health Organization. The ICGFI convened two task force meetings on irradiation as a quarantine treatment (Anonymous 1986a, 1991a). The ICGFI considered that tephritid fruit flies had been thoroughly researched and recommended that 150 Gy could be adopted as the minimum dose required to treat eggs or larvae to prevent emergence of normal adults for any species in the absence of specific data to support a lower dose. They adopted a generic disinfestation of 300 Gy as the minimum dose required to treat any preadult stage of insects to either prevent emergence of normal adults or sterilize any adults present and emerging from treated larvae or pupae. Also they recommended that, (1) the treatment schedules be incorporated in the International Plant Quarantine Treatment Manual (Stout 1983), (2) research be undertaken to develop treatment schedules for other pests of quarantine importance, (3) tolerance doses be evaluated for other commodities, (4) training programs be developed for plant protection and quarantine personnel, and (5) a technique be developed to identify irradiated insects. Several other meetings on this subject have been held in recent years. In 1982 a workshop on low dose irradiation treatment of agricultural commodities recommended that an interagency working group be formed to define, coordinate, and direct research, engineering, education, and liaison activities (McMullen & Yeager 1982). Also, irradiation as a quarantine treatment was discussed at an interAmerican meeting on harmonization of regulations related to trade in irradiated foods (Anonymous 1990a) and at a workshop on food irradiation (Anonymous 1990b).

Efficacy of Irradiation

Criteria for evaluating the effectiveness of treatments to eliminate fruit flies and other quarantine pests in the U.S. have been to prevent development of larvae in fruit and their successful pupation at the probit 9 (99.9968% mortality) security level (Baker 1939, Chew & Couey 1985, Couey & Chew 1986). Exposure of fruit fly larvae in water, fruit with a high juice content, or in an environment with a low oxygen level, may reduce the efficacy of an irradiation treatment and require an increase in dose to achieve quarantine security (Balock et al. 1963, Kaneshiro et al. 1983). A significant decrease in adult emergence when mature, cocooned, nondiapausing codling moth, *Cydia pomonella* (L.), larvae were irradiated resulted when the dose rate was increased from 1 to 200 Gy per min using a cobalt 60 source (Burditt et al. 1989). Although a minimal dose may be adequate for quarantine security, increasing the dose to 1,000 Gy could result in reduction of feeding or an increase in mortality of larvae, thus preventing further injury to the host commodity (Jona & Arzone 1979). Depending on dose, irradiation can result in mortality by preventing egg hatch, larval development, pupation, or adult emergence. Also, depending on dose, any adults developing from irradiated eggs or larvae possibly could not produce fertile offspring due to sterility or other abnormalities. Therefore, the criterion for efficacy of irradiation as a quarantine treatment could be based on preventing adult emergence (Balock et al. 1963) or on the inability of the insect to perpetuate the species (Ouye & Gilmore 1985).

Effect of Irradiation on Quarantine Pests

Diptera

Since 1960, numerous studies have been performed to determine the effects of irradiation on various quarantine pests. The objectives were to develop data to support proposed quarantine treatments and to perform basic research on the effects of irradiation on insect development. Most research on the effects of irradiation on quarantine pests has been conducted on Tephritidae. Results of research to determine the dose at which no adults emerged when fruit fly eggs or larvae were irradiated are summarized (Table 7.1). When lower doses were tested, many adults that emerged were abnormal and unable to reproduce. In some instances the listed dose was the lowest dose tested. Prevention of adult emergence and prevention of reproduction both have been proposed as

TABLE 7.1. Effects of exposure of fruit fly larvae to gamma radiation on prevention of subsequent adult emergence

Species	Age of Larvae (days)	Host	Dose (Gray) to Prevent Emergence (population tested)	
Anastrepha ludens	Mature	Grapefruit	50	(82)
Anastrepha suspensa	Mixed	Grapefruit	25	(1,285)
	Mixed	Grapefruit	154	(9,209)
	Mixed	Florida mango	17.5	(8,480)
	Mixed	Mango	25	(4,719)
	Mixed	Haitian mango	80	(2,961)
	Mixed	Carambola	50	(6,423)
	Mixed	Grapefruit	40	(3,808)
Bactrocera cucurbitae	Mixed	Fruit	100	(18,000)
Bactrocera dorsalis	Mixed	Mango	150	(197,041)[a]
Bactrocera jarvisi	5 d	Mango	101	(153,814)
Bactrocera tryoni	Young	Avocado	50	(20,373)
	Young	Orange	50	(9,915)
	Old	Avocado	75	(20,015)
	Old	Orange	75	(4,705)
	5 d	Mango	101	(138,635)
Ceratitis capitata	Mixed	Papaya	25	(19,000)
Rhagoletis indifferens	Mixed	Cherry	97	(84,369)

[a] One adult emerged.

criteria for security when irradiation is used as a quarantine treatment. As examples, two adult Mediterranean fruit flies emerged from papayas infested with eggs and larvae of different ages and treated at 100 Gy (Balock et al. 1966). No flies emerged from fruit treated at lower doses. Also, Seo et al. (1973) reported two Mediterranean fruit flies emerged from papayas infested with eggs and larvae of different ages treated at 225 Gy. Mediterranean fruit flies in these two reports emerged from papayas infested with melon flies and oriental fruit flies, indicating probable contamination of the fruit (Burditt 1982). Fésüs et al. (1981) found that when infested oranges were irradiated at 400 - 600 Gy, Mediterranean fruit fly eggs did not hatch, and larvae did not pupate. When fruits infested with Mediterranean fruit fly eggs were irradiated, the dose required to reduce egg hatch by 95% ranged from <10 Gy for young eggs in peaches to >600 Gy for older eggs in plums (Kaneshiro et al. 1983). Two adult oriental fruit flies emerged from papayas infested by eggs and larvae of different ages and treated at 100 Gy (Balock et al. 1966). Treatment in cartons at bulk densities of 0.30 - 0.42 g per cubic centimeter and 209 - 291 Gy of papayas infested with oriental fruit fly eggs and larvae resulted in emergence of 22 flies from an estimated population of >900,000 pupae (Seo et al. 1973). No oriental fruit flies emerged when larvae were irradiated at 250 Gy in mangoes (Loaharanu 1971). Thomas & Rahalkar et al. (1975) reported that irradiation of oriental fruit fly eggs and larvae at 150 or 250 Gy resulted in larvae that were sluggish, had a reduced growth rate, and no adults when eggs and larvae were treated at 150 Gy. They recommended 250 Gy for quarantine purposes, because they believed it would be effective against mango weevil and delay ripening of mangoes. Manoto & Blanco (1982) reported that when mangoes infested with oriental fruit fly were irradiated at 250 Gy 0, 2, 4, or 6 d after inoculation with eggs, adult emergence was zero, 0.1, 0.9, and 1%, respectively. In Thailand, Komson et al. (1988) irradiated 'Nang Klangwan' mangoes that had been infested with oriental fruit flies and held for 6 d prior to treatment. Recovery from the treated fruit was 145,912 pupae and one adult. No adults emerged from melon fly eggs or larvae irradiated in papayas, tomatoes, or cucumbers treated at 100 Gy (Balock et al. 1966). No adults emerged from bell peppers, eggplant, and papayas irradiated with 209 - 246 Gy at bulk densities of 0.23 - 0.42 g per centimeter (Seo et al. 1973). Thomas & Rahalkar (1975) reported that, similar to oriental fruit fly in mangoes, melon fly eggs or larvae irradiated in pumpkin at 150 or 250 Gy resulted in larvae that were sluggish and had a reduced growth rate. No adults emerged from eggs or larvae treated at 150 Gy. Nine adults emerged from 5,390 Queensland fruit fly pupae treated as larvae in apples and oranges at 40 Gy (Macfarlane 1966). Shipp & Osborn (1968) suggested 100 Gy to provide an adequate quarantine treatment to prevent spread of Queensland fruit fly if a delay in treatment occurred and pupae were present in cartons of fruit. Eric et al. (1970) found that 50 Gy would be an effective quarantine treatment to disinfest bananas. Doses of 25 - 50 Gy prevented emergence of adult Queensland fruit flies treated as young larvae in avocados and oranges, while 75 Gy was required for older larvae (Rigney & Wills 1985). When 5-d-old Queensland fruit fly and *Bactrocera jarvisi* (Tyron) larvae were treated in mangoes at 101 Gy, no adults emerged (Heather et al. 1991). Haque & Ahmad (1967) reported that no adults emerged from guavas infested with *Bactrocera zonata* (Saunders) eggs or early, intermediate, or full grown larvae irradiated at 45, 50, 50, or 55 Gy, respectively. Brownell & Yudelovitch (1962) irradiated grapefruit infested with Mexican fruit fly, *Anastrepha ludens* (Loew). They reported that no eggs or first instars treated at 20 Gy formed puparia, and no adults emerged from mature larvae treated at 50 Gy. When grapefruit infested with Caribbean fruit fly, *Anastrepha suspensa* (Loew), eggs or larvae was treated with a cesium 137 source, one adult emerged from fruit treated at 172 Gy, and a second emerged from fruit treated at 302 Gy (von Windeguth 1982). Both adults died before reaching sexual maturity. An abnormal adult emerged from infested grapefruit treated in a commercial cobalt 60 irradiator at 225 Gy (von Windeguth & Ismail 1987). No adults emerged from fruit treated at 40 or 80 Gy, with or without cold storage at 1.1°C for up to 6 d (von Windeguth & Gould 1990). von Windeguth (1986) treated mangoes infested with Caribbean fruit fly immatures in cobalt 60 irradiators at 5.1 or 218 Gy per min. At 5.1 Gy per min no adults emerged at 50 Gy, and at 10 or 25 Gy, many adults that emerged were abnormal. At 218 Gy per min, no adults emerged at 25 Gy. No adults emerged from carambolas treated at ≥50 Gy (Gould & von Windeguth 1991). No adults developed from European cherry fruit fly, *Rhagoletis cerasi* L., after larvae were exposed to 100 Gy in cherries (Jona & Arzone 1979). Larvae in fruit treated at 1,000 Gy did not form puparia, and 70% of those pupating after treatment at 300 or 500 Gy pupated in the fruit, compared with 1.2% of the controls. No normal adults developed from an estimated 84,368 larvae of western cherry fruit fly, *Rhagoletis indifferens* Curran, irradiated at 97 Gy in cherries (Burditt & Hungate 1988). One adult with vestigial wings emerged, and two adults failed to fully eclose from larvae that did not enter diapause.

Coleoptera

Irradiation has been researched on several curculionids. Johnson et al. (1990) showed that hatch of Fuller rose beetle, *Asynonychus godmani* Crotch, eggs 10 - 13 d old treated on lemons could be prevented by irradiation at 150 Gy. However, hatch of eggs 1 - 3 or 6 - 8 d old was prevented at 50 Gy. Sterilization of male khapra beetles, *Trogoderma granarium* Everts, was achieved at 250 - 300 Gy compared with doses as low as 70 Gy for legume weevils, *Callosobruchus* species, infesting pulses (Brower & Tilton 1985). Ignatowicz & Brzostek (1990) reported that exposing eggs or larvae to 90 and 60 Gy prevented reproduction of emerging adult bean weevils, *Acanthoscelides obtectus* Say, and rice weevils, *Sitophilus oryzae* (L.), respectively; whereas, 260 and 80 Gy prevented adult emergence. Seo et al. (1974) showed that 300 Gy reduced the development of immature stages and adult longevity of mango weevil. Milne et al. (1977) found that 500 Gy prevented emergence of adults.

Homoptera

Halfhill (1988) found that 80 Gy at 20°C prevented molting of fourth-instar asparagus aphids, *Brachycorynella asparagi* (Mordvilko), to adults and the production of progeny by adults infesting asparagus spears. Wit & van de Vrie (1985) reported that 100 Gy prevented the reproduction of green peach aphid, *Myzus persicae* (Sulzer). Angerilli & Fitzgibbon (1990) found that San Jose scale, *Quadraspidiotus perniciosus* (Comstock), survived on infested apples for 61 d in cold storage after treatment at 600 Gy.

Lepidoptera

Wit & van de Vrie (1985) showed that 200 Gy prevented adults developing from fifth instar leaf roller, *Clepsis spectrana* (Treitschke). Doses of 139 - 177 Gy prevented development of adults from 79,540 non-diapausing immature codling moth larvae infesting apples (Burditt & Hungate 1989). When larvae were treated at 58 - 98 Gy, significantly more adults emerged at 1 Gy per min than at 204 Gy per min (Burditt et al. 1989). No normal adults developed when 5,954 larvae were treated in walnuts at 177 Gy (Burditt 1986). Dentener et al. (1990) found that 199 Gy prevented development of larvae of light brown apple moth, *Epiphyas postvittana* (Walker), to the adult stage.

Thysanoptera

Wit & van de Vrie (1985) showed that for thrips, *Frankliniella pallida* Uzel, infesting cut flowers, one larva developed to the prepupa, and no larva developed to the adult at 100 Gy.

Acarina

Twospotted spider mite, *Tetranychus urticae* Koch, was sterilized at 350 Gy (Wit & van de Vrie 1985). Goodwin & Wellham (1990) found that 300 Gy disinfested cut flowers of all stages, and adults developing from treated juveniles were sterile. Ignatowicz & Brzostek (1990) showed that for mold mite, *Tyrophagus putrescentiae* (Schrank), and bulb mite, *Rhyizoglyphus echinopus* (Fumouze & Robin), adult sterility resulted from 260 and 300 Gy, respectively.

Identification of Irradiated Pests

A method to identify treated pests is needed to ensure that they have been irradiated. Rahman et al. (1990) reported that the area of the supraoesophageal ganglion was reduced in third instars developing from irradiated Mediterranean fruit fly eggs or larvae. There was a consistent reduction in the area of the supraoesophageal ganglion with an increase in radiation dose. The radiation sensitivity was distributed from egg through larval development. They proposed using this morphological change as a method for identifying treated fruit fly larvae and suggested that representative species of other genera of the family should be tested to formulate a generalized conclusion. This reduction in the ganglion could be more reliable than other indirect methods used earlier for identifying treated fruits (Chadwick et al. 1977).

Effect of Irradiation on Host Commodities

Generally, doses required for quarantine security are too low to control pathogens or to extend shelf life of host commodities. A dose of $\geq$5,000 Gy would be necessary to achieve pasteurization by irradiation (Moy 1983). Akamine & Moy (1983) reported that the shelf life of apricots, bananas, sweet cherries, figs, mangoes, papayas, strawberries, and tomatoes could be extended by irradiation at $\leq$1,000 Gy. Irradiation at $\leq$150 Gy inhibited sprouting of potatoes and onions (Matsuyama & Umeda 1983).

Research conducted for >30 years has shown that most fresh fruits and vegetables are not adversely affected by doses proposed for quarantine purposes (Abdel-Kader & Maxie 1967, Kader & Heintz 1983, Moy 1983, Akamine & Moy 1983). Kader (1986) listed commodities with high relative tolerance to irradiation at $\leq$1,000 Gy as logical choices for quarantine treatment (Table 7.2). This group included several commodities, such as melons, for which approved disinfestation methods are limited. Other commodities showed moderate or low tolerance at 1,000 Gy.

Sensory qualities of plums and nectarines irradiated at 300 Gy were comparable to those of the untreated controls (Moy et al. 1983, Moy & Nagai 1985). Differences were found in the color of peaches irradiated at 300 and 500 Gy and in flavor at 300 Gy. Differences in sensory qualities between irradiated samples and controls could be due to delayed ripening in irradiated fruits. The aroma and flavor of irradiated oranges were unaffected after treatment at $\leq$500 Gy. The ascorbic acid, total acidity, and total soluble solids of irradiated oranges and controls were not significantly different from each other. Ripening of 'Tommy Atkins' mangoes was delayed by 150 - 250 Gy, pH of the juice decreased, and titratable acidity increased as dose increased. Internal breakdown increased at $\geq$250 Gy (Spalding & von Windeguth 1988). No phytotoxicity was observed when 'Arkin' carambolas were treated at $\leq$600 Gy (Gould & von Windeguth 1991).

TABLE 7.2. Relative tolerances of fresh fruits and vegetables to irradiation stress at doses ≤1 kGy [a]

Relative Tolerance		
High	Moderate	Low
Apple	Apricot	Avocado
Cherry	Banana	Broccoli
Date	Cherimoya	Cauliflower
Guava	Fig	Cucumber
Longan	Grapefruit	Grape
Mango	Kumquat	Green bean
Muskmelon	Loquat	Leafy vegetables
Nectarine	Lychee	Lemon
Papaya	Orange	Lime
Peach	Passion fruit	Olive
Rambutan	Pear	Pepper
Raspberry	Pineapple	Sapodilla
Strawberry	Plum	Soursop
Tamarillo	Tangelo	Summer squash
Tomato	Tangerine	

[a] Adapted from Kader (1986).

Most studies have involved the effects of irradiation from gamma sources on host commodities. The present trend to develop linear accelerators would require extensive studies to determine if the high dose rates produced by such machines would result in adverse affects on the commodities.

Conclusions

Irradiation is a viable alternative treatment for pests of quarantine importance. However, industry and regulatory officials must consider its benefits in relation to other alternatives and the treatment with the best benefit/cost ratio. If irradiation is accepted, industry and the public must be convinced that the product can be irradiated economically and without damage. Education is needed to raise the level of understanding of the process, its use as a quarantine treatment, and to improve public acceptance of irradiated products and irradiation facilities. Short courses for training plant quarantine and other agricultural industry personnel should be conducted. This process could help identify problems and concerns, and increase the understanding of irradiation technology by the public.

Although electron beam and X ray sources have been considered suitable for treating commodities for quarantine purposes, limited research has been done to demonstrate their efficacy. Research is needed to develop mortality data and quarantine security doses, and to determine the phytotoxicity or other adverse effects on treated commodities from these sources.

References

Abdel-Kader, A. S. & E. C. Maxie. 1967. "Radiation Pasteurization of Fruits and Vegetables: a Bibliography," in P. S. Baker, ed., *Isotopes Information Center. IIC-11*. Pp. 1-52. Oak Ridge, Tennessee: Oak Ridge National Library.

Akamine, E. K. & J. H. Moy. 1983. "Delay in Postharvest Ripening and Senescence of Fruits," in E. S. Josephson & M. S. Peterson, eds., *Preservation of Food by Ionizing Radiation, Vol. III*. Pp. 129-158. Boca Raton, Florida: CRC Press.

Angerilli, N.P.D. & F. Fitzgibbon. 1990. Effects of cobalt gamma radiation on San Jose scale (Homoptera: Diaspididae) survival on apples in cold and controlled atmosphere storage. J. Econ. Entomol. 83: 892-895.

Anonymous. 1971. "Disinfestation of Fruit by Irradiation." *Proceedings of a Panel on the Use of Irradiation to Solve Quarantine Problems in the International Fruit Trade, Honolulu 1970*. 177 pp. Vienna, Austria: International Atomic Energy Agency.

_____. 1981. "Wholesomeness of Irradiated Foods." *Report of the Joint FAO/IAEA/WHO Expert Committee*. Technical Report Series No. 659. WHO, Geneva.

_____. 1984a. Food and Drug Administration. Irradiation in the production processing and handling of food. Federal Register 49: 5714-5722.

_____. 1984b. "Report on FAO/IAEA Consultant Group Meeting on the Use of Irradiation as a Quarantine Treatment of Agricultural Commodities." *Food Irradiation Newsletter* 8: 6-13.

_____. 1984c. "Report of the FAO/IAEA Research Coordination Meeting on Insect Disinfestation of Food and Agricultural Products by Irradiation." *Food Irradiation Newsletter* 8: 14-28.

_____. 1985. "Report of the Second FAO/IAEA Research Coordination Meeting on Insect Disinfestation of Food and Agricultural Products by Irradiation." *Food Irradiation Newsletter* 9: 12-28.

_____. 1986a. "Excerpt of the Task Force Meeting on Irradiation as a Quarantine Treatment." Chiang Mai, Thailand. 17-21 February 1986.

Food Irradiation Newsletter 10: 5-10.

_____. 1986b. "Report of the FAO/IAEA Research Coordination Meeting on the Use of Irradiation as a Quarantine Treatment of Food and Agricultural Commodities." Chiang Mai, Thailand. 24-28 February 1986. *Food Irradiation Newsletter* 10: 17-28.

_____. 1989a. Department of Agriculture, Animal and Plant Health Inspection Service. Use of irradiation as a quarantine treatment for fresh fruits of papaya from Hawaii. Federal Register 54: 387-393.

_____. 1989b. "Report of the Second FAO/IAEA Research Co-ordination Meeting on the Use of Irradiation as a Quarantine Treatment of Food and Agricultural Commodities." Orlando, Florida. 23-27 May 1988. *Food Irradiation Newsletter* 13: 34-48.

_____. 1990a. "Inter-American Meeting on Harmonization of Regulations Related to Trade in Irradiated Foods." Orlando, Florida U.S.A. 27 November to 1 December 1990. *Food Irradiation Newsletter* 14: 12-21.

_____. 1990b. "RCA Workshop on Commercialization of Food Irradiation." Shanghai, People's Republic of China. 8-12 January 1990. *Food Irradiation Newsletter* 14: 11-23.

_____. 1991a. "Summary of Task Force Meeting on Irradiation as a Quarantine Treatment of Fresh Fruits and Vegetables." Bethesda, Maryland. 7-11 January 1991. *Food Irradiation Newsletter* 15: 9-10.

_____. 1991b. "Report of the Final FAO/IAEA Research Co-ordination Meeting on the Use of Irradiation as a Quarantine Treatment of Food and Agricultural Commodities." Malaysian Agricultural Research and Development Institute (MARDI), Kuala Lumpur, Malaysia. 27-31 August 1990. *Food Irradiation Newsletter* 15: 42-58.

Baker, A. C. 1939. The basis for treatment of products where fruitflies are involved as a condition for entry into the United States. USDA Circular 551.

Balock, J. W., L. D. Christenson & G. O. Burr. 1956. "Effect of Gamma Rays from Cobalt 60 on Immature Stages of the Oriental Fruit Fly (*Dacus dorsalis* Hendel) and Possible Application to Commodity Treatment Problems" (Abstract), in *Proceedings of the 31st Annual Meeting of the Hawaii Academy of Sciences, Honolulu.* P. 18. Honolulu: Hawaii Academy of Sciences.

Balock, J. W., A. K. Burditt, Jr., & L. D. Christenson. 1963. Effects of gamma radiation on various stages of three fruit fly species. J. Econ. Entomol. 56: 42-46.

Balock, J. W., A. K. Burditt, Jr., S. T. Seo & E. K. Akamine. 1966. Gamma radiation as a quarantine treatment for Hawaiian fruit flies. J. Econ. Entomol. 59: 202-204.

Brower, J. H. & E. W. Tilton. 1985. "The Potential of Irradiation Treatment for Insects Infesting Stored-Food Commodities," in J. H. Moy, ed., *Radiation Disinfestation of Food and Agricultural Products, Proceedings of an International Conference, Honolulu. November 14-18, 1983.* Pp. 75-86. Honolulu, Hawaii: Hawaii Institute of Tropical Agr. and Human Resources. University of Hawaii at Manoa.

Brownell, L. E. & M. Yudelovitch. 1962. "Effect of Radiation on Mexican Fruit-Fly Eggs and Larvae in Grapefruit," in *Proceedings, Symposium Radioisotopes and Radiation in Entomology.* Pp. 193-202. Bombay, India. December 5-9, 1960. Vienna, Austria: International Atomic Energy Agency.

Burditt, A. K., Jr. 1982. Food irradiation as a quarantine treatment of fruits. Food Technology 36: 51-54, 58-60, 62.

_____. 1986. Gamma irradiation as a quarantine treatment for walnuts infested with codling moths (Lepidoptera: Tortricidae). J. Econ. Entomol. 79: 1577-1579.

Burditt, A. K., Jr. & F. P. Hungate. 1988. Gamma irradiation as a quarantine treatment for cherries infested by western cherry fruit fly (Diptera: Tephritidae). J. Econ. Entomol. 81: 859-862.

_____. 1989. Gamma irradiation as a quarantine treatment for apples infested by codling moth (Lepidoptera: Tortricidae). J. Econ. Entomol. 82: 1386-1390.

Burditt, A. K., Jr., F. P. Hungate & H. H. Toba. 1989. Gamma Irradiation: Effect of dose rate on development of mature codling moth larvae and adult eclosion. Radiation Physics and Chemistry 34: 979-984.

Chadwick, K. H., D.A.E. Ehlermann & W. L. McLaughlin. 1977. *Manual of Food Irradiation Dosimetry, Technical Reports Series 178.* Vienna, Austria: International Atomic Energy Agency.

Chew, V. & M. T. Ouye. 1985. "Statistical Basis for Quarantine Treatment Schedule and Security," in J. H. Moy, ed., *Radiation Disinfestation of Food and Agricultural Products, Proceedings of an International Conference, Honolulu (1983).* Pp. 70-74. Honolulu, Hawaii: Hawaii Institute of Tropical Agr. and Human Resources. University of Hawaii at Manoa, Honolulu.

Couey, H. M. & V. Chew. 1986. Confidence limits and sample size in quarantine research. J. Econ. Entomol. 79: 887-890.

Dentener, P. R., B. C. Waddell & T. A. Batchelor. 1990. "Disinfestation of Light Brown Apple Moth: a Discussion of Three Disinfestation Methods," in B. B. Beattie, ed., *Managing Postharvest Horticulture in Australasia, Proceedings of The Australasian Conference in Postharvest Horticulture, 24-28 July 1989, Gosford, NSW, Australia..* Pp. 166-177. Australian Institute of Agric. Sci., Sydney, Australia. Occasional

114

Publication No. 46.

Eric, B., J. LeCompte, S. Klein & W. Kricker. 1970. Study of disinfestation of bananas by gamma irradiation. Food Technology in Australia 22: 664-667.

Fésüs, I., L. Kádas, & B. Kálmán. 1981. Protection of oranges by gamma radiation against *Ceratitis capitata* Wied. Acta Alimentaria 10: 293-299.

Goodwin, S. & T. M. Wellham. 1990. Gamma irradiation for disinfestation of cut flowers infested by twospotted spider mite (Acarina: Tetranychidae). J. Econ. Entomol. 83: 1455-1458.

Gould, W. P. & D. L. von Windeguth. 1991. Gamma irradiation as a quarantine treatment for carambolas infested with Caribbean fruit flies. Fla. Entomologist 74: 297-300.

Halfhill, J. E. 1988. Irradiation disinfestation of asparagus spears contaminated with *Brachycorynella asparagi* (Mordvilko) (Homoptera: Aphididae). J. Econ. Entomol. 81: 873-876.

Haque, H. & R. Ahmad. 1967. Effect of ionizing radiation on *Dacus zonatus* fruit fly eggs and larvae *in situ*. Pakistan J. Sci. 19: 233-238.

Heather, N. W., R. J. Corcoran & O. Banos. 1991. Disinfestation of mangoes with gamma irradiation against two Australian fruit flies (Diptera: Tephritidae). J. Econ. Entomol. 84: 1304-1307.

Ignatowicz, S. & G. Brzostek. 1990. Use of irradiation as a quarantine treatment for agricultural products infested by mites and insects. Radiation Physics and Chemistry 35: 263-267.

Johnson, J. A., E. L. Soderstrom, D. G. Brandl, G. Houck & P. L. Wofford. 1990. Gamma radiation as a quarantine treatment for Fuller rose beetle eggs (Coleoptera: Curculionidae) on citrus fruit. J. Econ. Entomol. 83: 905-909.

Jona, R. & A. Arzone. 1979. Control of *Rhagoletis cerasi* in cherries by gamma irradiation. J. Hortic. Sci. 54: 167-170.

Kader, A. A. 1986. Potential applications of ionizing radiation in postharvest handling of fresh fruits and vegetables. Food Technology 40: 117-121.

Kader, A. A. & C. M. Heintz. 1983. Gamma Irradiation of Fresh Fruits and Vegetables, an Indexed Reference List (1965-1982). Pp. 1-55. Davis, California: University of California.

Kaneshiro, K. Y., A. T. Ohta, J. S. Kurihara, K. M. Kanegawa & L. R. Nagamine. 1983. Gamma radiation treatment for disinfestation of the Mediterranean fruit fly in California grown fruits. I. Stone fruits. Proceedings, Hawaiian Entomological Soc. 24: 245-259.

Koidsumi, K. 1930. Quantitative studies on the lethal action of x-rays upon certain insects. J. of the Soc. of Tropical Agr. 2: 243-263.

Komson, P., M. Sutantawong, E. Smitasiri, C. Lapasatukul & U.

Unahawutti. 1988. "Irradiation as a Quarantine Treatment for the Oriental Fruit Fly, *Dacus dorsalis* Hendel, in Mangoes," *Proceedings of an International Symposium on Modern Insect Control: Nuclear Techniques and Biotechnology.* Pp. 319-324. Vienna, Austria: International Atomic Energy Agency.

Loaharanu, P. 1971. "Recent Research on the Influence of Irradiation of Certain Tropical Fruits in Thailand," in *Disinfestation of Fruit by Irradiation, Proceedings of a Panel on the Use of Irradiation to Solve Quarantine Problems in the International Fruit Trade, Honolulu 1970.* Pp. 113-124. Vienna, Austria: International Atomic Energy Agency.

Macfarlane, J. J. 1966. Control of the Queensland fruit fly by gamma irradiation. J. Econ. Entomol. 59: 884-889.

Manoto, E. C. & L. R. Blanco. 1982. Disinfestation of mangoes by gamma irradiation. J. Radioisotope Soc. Philippines 20: 52-57.

Matsuyama, A. & K. Umeda. 1983. "Sprout Inhibition in Tubers and Bulbs," in E. S. Josephson & M. S. Peterson, eds., *Preservation of Food by Ionizing Radiation, Vol. III.* Pp. 159-213. Boca Raton, Florida: CRC Press.

McMullen, W. H. & J. G. Yeager. 1982. "Workshop on Low-Dose Radiation Treatment of Agricultural Commodities," April 19-21, 1982, Working Report. 74 pp. Department of Energy/United States Department of Agr./American Institute of Biological Sciences.

Milne, D. L., I. B. Kok, A. C. Thomas & D. H. Swarts. 1977. Inactivation of mango seed weevil, *Sternochetus mangiferae*, by gamma irradiation. Citrus and Subtropical Fruit J. (Jan. 1977.) Pp. 11-17.

Moy, J. H. 1983. "Radurization and Radicidation: Fruits and Vegetables," in E. S. Josephson & M. S. Peterson, eds., *Preservation of Food by Ionizing Radiation, Vol. III.* Pp. 83-108. Boca Raton, Florida: CRC Press.

Moy, J. H. & N. Y. Nagai. 1985. "Quality of Fresh Fruits Irradiated at Disinfestation Doses," in J. H. Moy, ed., *Radiation Disinfestation of Food and Agricultural Products, Proceedings of an International Conference, Honolulu. November 14-18, 1983.* Pp. 135-145. Honolulu, Hawaii: Hawaii Institute of Tropical Agr. and Human Resources. University of Hawaii at Manoa.

Moy, J. H., K. Y. Kaneshiro, A. T. Ohta & N. Nagai. 1983. Radiation disinfestation of California stone fruits infested by Medfly -- Effectiveness and fruit quality. J. Food Sci. 48: 928-931, 934.

Ouye, M. T. & J. E. Gilmore. 1985. "The Philosophy of Quarantine Treatment as Related to Low-Dose Radiation," in J. H. Moy, ed., *Radiation Disinfestation of Food and Agricultural Products, Proceedings of an International Conference, Honolulu. November 14-18, 1983.* Pp. 67-69. Honolulu, Hawaii: Hawaii Institute of Tropical Agr. and Human

Resources. University of Hawaii at Manoa.

Prusik, T. & T. Wallace. 1985. "An Automated System for Measuring the Dose Provided to Irradiated Food," in *Food Irradiation Processing, Proceedings of an International Symposium on Food Irradiation Processing, Washington, D.C. March 4-8, 1985*. Vienna, Austria: International Atomic Energy Agency.

Rahman, R., C. Rigney & E. Busch-Petersen. 1990. Irradiation as a quarantine treatment against *Ceratitis capitata* (Diptera: Tephritidae): Anatomical and cytogenetic changes in mature larvae after gamma irradiation. J. Econ. Entomol. 83: 1449-1454.

Rigney, C. J. & P. A. Wills. 1985. "Efficacy of Gamma Irradiation as a Quarantine Treatment Against Queensland Fruit Fly," in J. H. Moy, ed., *Radiation Disinfestation of Food and Agricultural Products, Proceedings of an International Conference, Honolulu. November 14-18, 1983*. Pp. 116-120. Honolulu, Hawaii: Hawaii Institute of Tropical Agr. and Human Resources. University of Hawaii at Manoa.

Seo, S. T., R. M. Kobayashi, D. L. Chambers, A. M. Dollar & M. Hanaoka. 1973. Hawaiian fruit flies in papaya, bell pepper, and egg-plant: quarantine treatment with gamma irradiation. J. Econ. Entomol. 66: 937-939.

Seo, S. T., R. M. Kobayashi, D. L. Chambers, L. F. Steiner, C.Y.L. Lee & M. Komura. 1974. Mango weevil: Cobalt-60 gamma irradiation of packaged mangoes. J. Econ. Entomol. 67: 504-505.

Shipp, E. & A. W. Osborn. 1968. Irradiation of Queensland fruit fly pupae to meet quarantine requirements. J. Econ. Entomol. 61: 1721-1726.

Spalding, D. H. & D. L. von Windeguth. 1988. Quality and decay of irradiated mangoes. HortScience 23: 187-189.

Stout, O. O. (Revised by H. L. Roth). 1983. In J. F. Karpati, C. Y. Schotman & K. A. Zammarano, eds., *International Plant Quarantine Treatment Manual. FAO Plant Production and Protection Paper 50*. Pp. x+220. Rome, Italy: Food and Agr. Organization of the United Nations.

Thomas, P. & G. W. Rahalkar. 1975. Disinfestation of fruit flies in mango by gamma irradiation. Current Sci. 44: 775-776.

von Windeguth, D. L. 1982. Effects of gamma irradiation on the mortality of the Caribbean fruit fly in grapefruit. Proceedings, Fla. State Hortic. Soc. 95: 235-237.

_____. 1986. Gamma irradiation as a quarantine treatment for Caribbean fruit fly infested mangoes. Proceedings, Fla. State Hortic. Soc. 99: 131-134.

von Windeguth, D. L. & W. P. Gould. 1990. Gamma irradiation followed by cold storage as a quarantine treatment for Florida grapefruit infested with Caribbean fruit fly. Fla. Entomologist 73: 242-247.

von Windeguth, D. L. & M. A. Ismail. 1987. Gamma irradiation as a quarantine treatment for Florida grapefruit infested with Caribbean fruit fly, *Anastrepha suspensa* (Loew). Proceedings, Fla. State Hortic. Soc. 100: 5-7.

Wit, A.K.H. & M. van de Vrie. 1985. Gamma radiation for post harvest control of insects and mites in cut flowers. Mededelingen van de Faculteit Landbouwwetenschappen, Rijksuniversiteit Gent 50/2b, 697-704.

8

Cold Storage

Walter P. Gould

Cold storage was originally used to inhibit decay and extend the shelf life of commodities. Modern cold storage facilities and refrigerated transportation technology enabled the fresh fruit industry, which was limited and regional prior to 1800, to expand substantially. Widespread marketing of food products also brought a wide dispersal of some major economic pests. This resulted in the subsequent formulation of quarantine regulations designed to restrict the spread of pests. Mechanical refrigeration made possible the use of cold temperature against insect pests.

Use of Cold Storage Against Insects

Research on the use of cold to destroy insect pests began in the 1890s and expanded to cover a variety of commodity pests early in this century. Cold (from −7.7 to 8.8°C from a few days to several months) was tested as a method to control webbing clothes moth, *Tineola biselliella* (Hummel), black carpet beetle, *Attagenus unicolor* (Brahm), *Dermestes vulpinus* F., *Tenebrio obscurus* F., and cabinet beetle, *Trogoderma tarsalis* (Melsh), in stored goods (Howard 1896). Duvel (1905) found that 0 - 1°C could be used effectively to prevent infestation of stored cowpeas by cowpea weevil, *Callosobruchus maculatus* F., and four-spotted bean weevil, *Bruchus quadrimaculatus* F.

Mediterranean fruit fly, *Ceratitis capitata* (Wiedemann), became established in many parts of the world early in this century from a suspected origin in East Africa (Enkerlin et al. 1989). Fuller (1906) reported the survival of Mediterranean fruit flies from a shipment of peaches kept at 3.9 - 4.4°C for 124 d. Lounsbury (1907), however, reported that 3.3 - 4.4°C killed all larvae in three weeks. Hooper (1907) reported that the flies survived 3.3 - 4.4°C but were killed in 15 d at 0.56 - 1.67°C. Wilcox & Hunn (1914) reported that Mediterranean fruit fly larvae did not survive 0°C for 4.5 d, and adults failed to emerge from larvae held for 2.5 d at this temperature.

Back & Pemberton (1916a, 1916b, 1918) did the first extensive laboratory work on the effect of cold temperatures on eggs, larvae, and pupae of Mediterranean fruit fly. They found that 4.4 - 7.2°C killed all stages in seven weeks, 0.56 - 4.4°C killed all stages in three weeks, 0 - 0.56°C killed all stages in two weeks, and -4.4 to -1.1°C killed eggs in 7 d, larvae in 6 d and pupae in 4 d. In South Africa, Mediterranean fruit fly larvae survived from -3.9 to 6.7°C for six weeks (Union of South Africa Department of Agriculture 1923). The temperature range for the South African work was reported as fluctuating from -3.9 to 6.7°C. Pettey & Griffiths (1931) reported no stages of Mediterranean fruit fly survived three weeks at 0 ± 0.8°C. Mason & McBride (1934) reported that complete mortality of all immature stages required 8 - 11 d at -1.67 to -0.56°C under commercial storage conditions in Hawaii. Nel (1936) reported that 9 d at -0.56°C, 12 d at 1.1°C, and 16 d at 2.77°C destroyed all stages of Mediterranean fruit fly.

The main application of this early work was the justification of continued shipping of fruits, using commercial cold storage, from regions where Mediterranean fruit fly was present. The first use of cold solely as a quarantine treatment was in the United States (U.S.) against the first Mediterranean fruit fly outbreak in Florida in 1929 (Baker 1952, Richardson 1952, 1958). A cold treatment of -2.2°C for 5 h followed by -1.1°C for 5 d was used but later changed to 1.1°C for 12 d after the former treatment was found to damage citrus (Richardson 1958). Cold treatment from -1.1 to -0.55°C for 15 d was used for citrus from Texas and Mexico infested with Mexican fruit fly, *Anastrepha ludens* (Loew), from 1929 to 1937 (Baker et al. 1944).

Temperatures from -1.1 to 7.2°C were studied for use against oriental fruit fly, *Bactrocera dorsalis* (Hendel), and melon fly, *Bactrocera cucurbitae* (Coquillett), in Hawaii (Joint Legislative Committee on Agriculture and Livestock Problems 1953, Burditt & Balock 1985). The time to kill all stages of oriental fruit fly was 9 d at -1.1°C, 12 d at 2.7°C, 14 d at 4.4°C, and >28 d at 7.2°C. All stages of melon fly were killed after 7 d at -1.1°C and 10 d at 2.7°C. Temperatures <7.2°C damaged most Hawaiian fruits and vegetables.

From 1935 through 1938, tests were conducted in Puerto Rico on mangoes and guavas infested with West Indian fruit fly, *Anastrepha obliqua* (Macquart), and Caribbean fruit fly, *Anastrepha suspensa* (Loew), at 0.0 - 2.2°C (Burditt & McAlister 1982). No larvae survived 9 d at 0°C, 13 d at 1.1°C, and 16 d at 2.2°C. Mangoes and guavas did not tolerate these temperatures; hence, cold treatments were not commercially used.

These studies have led to the development of a protocol for using in-transit cold storage as a commodity treatment for potentially infested fruit. The United States Department of Agriculture, Animal and Plant Health Inspection Service, Plant Protection and Quarantine (USDA-APHIS-PPQ) Treatment Manual lists times and temperatures (Table 8.1) required for cold treatment of fruits and vegetables, according to their country of origin (Table 8.2) (Anonymous 1992a). Treatment procedures and facilities must meet certain requirements defined by the USDA-APHIS-PPQ. The cold treatment schedule T107 of the Treatment Manual lists the protocol for Mediterranean fruit fly, Mexican fruit fly, other species of *Anastrepha*, Queensland fruit fly, *Bactrocera tryoni* (Froggatt), and false codling moth, *Cryptophlebia leucotreta* (Meyrick). The listed fruits are mostly temperate or citrus fruits coming from Central and South America (*Anastrepha* species and Mediterranean fruit fly), Europe and South Africa (Mediterranean fruit fly), and Australia (Queensland fruit fly). All treatments are for fruit destined for the U.S. Recently, apples from Guyana, lychees from Israel, and apples, kiwifruits, and pears from Zimbabwe have been added to the USDA-APHIS-PPQ cold treatment schedules (Clayton 1993).

Within the U.S. some states have quarantines. A 40 d, −0.6°C quarantine treatment for plum curculio, *Conotrachelus nenuphar* (Herbst), apple maggot, *Rhagoletis pomonella* (Walsh), and blueberry maggot, *Rhagoletis mendax* Curran, in apple, apricot, blueberry, cherry, hawthorne, huckleberry, nectarine, peach, pear, plum, prune, and quince is required for importation to California (Stout & Roth 1983). Arizona regulations require a treatment of 90 d at ≥3.3°C against plum curculio in apples (Stout & Roth 1983).

A 42 d at −0.6 to 2.2°C quarantine treatment for apple maggot in apples is required by Canada (Stout & Roth 1983). In Western Australia, cold storage of 0.5 ± 0.5°C for 14 d and 1.5 ± 0.5°C for 16 d was studied as a potential method to disinfest apples of Mediterranean fruit fly as part of the normal storage and handling procedures (Sproul 1976). Cold treatments of 1 ± 0.5°C for 12 d and 16 d have been studied as quarantine treatments for kiwifruit and table grapes infested with Queensland fruit fly and oranges infested with Queensland fruit fly and Mediterranean fruit fly in Australia, respectively (Hill et al. 1988, Jessup & Baheer 1990, Jessup 1992, Delima 1992).

With the banning of the fumigant ethylene dibromide (Ruckelshaus 1984), research on other types of commodity treatments became a

TABLE 8.1 USDA-APHIS-PPQ Treatment Manual, pages 5.58 - 5.59, listing schedules for quarantine cold treatments currently authorized

Section T107 - <u>Cold treatment</u>

(a) For *Ceratitis capitata* . . . . . 10 days at 0°C (32°F) or below
11 days at 0.55°C (33°F) or below
12 days at 1.11°C (34°F) or below
14 days at 1.66°C (35°F) or below
16 days at 2.22°C (36°F) or below

Alternate treatment: fumigation plus refrigeration T108

(b) For *Anastrepha ludens* . . . . 18 days at 0.55°C (33°F) or below
20 days at 1.11°C (34°F) or below
22 days at 1.66°C (35°F) or below

(c) For other species of *Anastrepha* . . . .

11 days at 0°C (32°F) or below
13 days at 0.55°C (33°F) or below
15 days at 1.11°C (34°F) or below
17 days at 1.66°C (35°F) or below

(d) For *Bactrocera tryoni* . . . . . 13 days at 0°C (32°F) or below
14 days at 0.55°C (33°F) or below

18 days at 1.11°C (34°F) or below
20 days at 1.66°C (35°F) or below
22 days at 2.22°C (36°F) or below

(e) For *Cryptophlebia leucotreta* . . . 22 days at -0.55°C (31°F) or below

If the temperature exceeds -0.27°C (31.5°F), the treatment shall be extended 1/3 of a day for each day or part of a day the temperature is above -0.27°C (31.5°F). If the temperature exceeds 1.11°C (34°F) at any time, the treatment is nullified.

Notes: <u>Pulp of the Fruit</u> must be at or below the indicated temperature at time of beginning treatment (for all T107 treatments).

(Rev. November 1992)

TABLE 8.2 USDA-APHIS-PPQ Treatment Manual, pages 5.56 - 5.58, listing fruits, pests and countries of origin for which cold treatments are authorized (revised November 1992)

Fruits for which cold treatment is authorized from specified countries of origin. Use schedule indicated by the letter following the country name. More than one pest may be involved.

Albania	(a)	ethrog
Algeria	(a)	ethrog, grape*, grapefruit, pear, plum, orange, tangerine
Argentina	(c)	apple, apricot, cherry, grape, nectarine, peach, pear, plum, pomegranate, quince
Armenia	(a)	grape*
Australia	(d)	apple, pear, kiwi
Austria	(a)	grape*
Azerbaijan	(a)	grape*
Belize	(b)	grapefruit, tangerine
Bermuda	(a)	grapefruit, orange
Bolivia	(c)	grapefruit, orange
Bosnia	(a)	ethrog
Brazil	(c)	grape
Bulgaria	(a)	grape*
Byelorus	(a)	grape*
Chile	(a)	apple, apricot**, cherry, grape**, kiwi, nectarine**, peach**, pear, plum**, quince
Colombia	(b)	grapefruit, orange, plum, tangerine
Colombia	(c)	grape
Corsica	(a)	ethrog
Costa Rica	(b)	ethrog, grapefruit, orange, tangerine
Croatia	(a)	ethrog
Cyprus	(a)	ethrog, grape*
Dominican Republic	(c)	grape
Ecuador	(c)	grape
Ecuador	(a)	ethrog
Egypt	(a)	grape*, orange, pear
El Salvador	(b)	ethrog, grapefruit, orange, tangerine
France	(a)	apple, ethrog, grape*, pear
Germany	(a)	grape*
Greece	(a)	ethrog, grape*, kiwi, orange, tangerine
Guatemala	(b)	ethrog, grapefruit, orange, plum, tangerine
Guyana	(c)	orange
Haiti	(c)	pomegranate
Honduras	(b)	ethrog, grapefruit, orange, tangerine

(Continues)

TABLE 8.2 *(Continued)*

<u>Fruits for which cold treatment</u> is authorized from specified countries of origin. Use schedule indicated by the letter following the country name. More than one pest may be involved.

Hungary	(a)	apple, grape*
Israel	(a)	apple, ethrog, grape*, grapefruit, loquat, orange, plum, pomegranate, tangerine
Italy	(a)	apple, ethrog, grape*, grapefruit, orange, pear, persimmon, tangerine
Jordan	(a)	apple, persimmon
Kazakhstan	(a)	grape*
Kirghiz	(a)	grape*
Latvia	(a)	grape*
Lebanon	(a)	apple
Libya	(a)	ethrog, grape*
Lithuania	(a)	grape*
Luxembourg	(a)	grape*
Macedonia	(a)	ethrog
Maldavia	(a)	grape*
Mexico	(b)	apple, grapefruit, orange, plum, tangerine
Montenegro	(a)	ethrog
Morocco	(a)	apricot, ethrog, grape*, grapefruit, orange, peach, pear, plum, tangerine
Nicaragua	(b)	ethrog, grapefruit, orange, tangerine
Panama	(b)	ethrog, grapefruit, orange, tangerine
Peru	(c)	grape
Portugal	(a)	ethrog, grape*, pear, apple
Rep. of South Africa	(a)	apple, apricot, grape, nectarine, passion fruit, peach, pear, plum
Russia	(a)	grape*
Serbia	(a)	ethrog
Spain	(a)	ethrog, grape*, grapefruit, orange, pear, tangerine
Suriname	(c)	grapefruit, orange, tangerine
Switzerland	(a)	grape*
Syria	(a)	ethrog, grape*
Tadzhikistan	(a)	grape*
Tunisia	(a)	ethrog, grape*, grapefruit, orange, peach, pear, plum, tangerine
Turkey	(a)	ethrog
Turkmenistan	(a)	grape*
Ukrania	(a)	grape*

TABLE 8.2 *(Continued)*

<u>Fruits for which cold treatment</u> is authorized from specified countries of origin. Use schedule indicated by the letter following the country name. More than one pest may be involved.

Uruguay	(c)	apple, grape, nectarine, peach, pear
Uzebekistan	(a)	grape*
Venezuela	(c)	grapefruit, orange, tangerine

* T101(i^2) also required.
** T101(a^3) also required.

Treatment upon arrival may be accomplished at northern ports as named in the permits; treatment in transit may be authorized for specifically equipped and approved vessels and from approved countries, for entry at ports named in the permits. Intransit cold treatment authorization must be preceded by a visit to the country of origin by a PPQ official to explain loading, inspection, and certification procedures to designated certifying officials of country of origin. Refrigerated compartments on carrying vessels and cold storage warehouse must have prior certification by PPQ. Authorization of cold treatments from countries with direct sailing time less than the number of days prescribed for intransit refrigeration treatment must be contingent on importer understanding that prescribed intransit refrigeration period must be met before arrival of vessel at the approved U.S. port.

(Rev. November 1992)

---

necessity. Cold treatment of 1.1°C for 21 d is used to kill Caribbean fruit fly larvae in citrus destined for Japan (Benschoter 1979, 1981, 1983, 1984, 1987, Benschoter & Witherell 1984). Gould & Sharp (1990) developed a cold treatment of 1.1°C for 15 d for Florida carambolas shipped to California.

Cold has shown promise as a quarantine treatment against codling moth, *Cydia pomonella* (L.), in apples and stone fruits. Dustan (1963) found that 0 ± 0.56°C for 56 d provided 100% mortality of larval stages of codling moth infesting apples. Moffit & Albano (1972) found that -0.56 ± 0.28°C for 133 d killed 100% of all codling moth stages except diapausing larvae. Moffit & Burditt (1989) found that 36 - 42 d exposure to 0°C gave 100% mortality of codling moth eggs. Yokoyama & Miller (1989) found that 0°C for 21 d killed 100% of codling moth infesting stone fruits but not oriental fruit moth, *Grapholita molesta* (Busck).

Mangosteens infested with oriental fruit fly were predicted to require treatment for 19.5 d at 5°C, 24.9 d at 6°C, and 24.9 d at 7°C to achieve probit 9 quarantine security (99.9968% mortality) (Burikam et al. 1992). These temperatures are warmer than those that have been used against other pests, but were effective against tropical fruit flies. Gould & Sharp (1990) found that 22 d at 5°C did not achieve quarantine security for Caribbean fruit fly in carambola, presumably because this fly is adapted to colder conditions. Chen et al. (1990) found that tropical flesh flies (Sarcophagidae) were more susceptible to both heat and cold than their temperate relatives. Davidson (1990) found a similar pattern with *Drosophila* species.

Cold is not a viable treatment for many temperate insects that are adapted to cold (Baust & Rojas 1985). Meats (1976) found that Queensland fruit fly acclimated to cold temperatures. Czajka & Lee (1990) found that *Drosophila* could acclimate (30 min at 5°C) so that they could survive several hours at temperatures which are normally lethal.

Quick freezing at −17°C is an approved quarantine treatment (Anonymous 1992b). Materials other than fruit pulp or slices generally do not tolerate the treatment without showing major damage; however, culled apples were quick frozen and sent from the U.S. to Japan for processing into juice (H. Moffitt, unpublished data).

Advantages and Problems of Cold as a Quarantine Treatment

Cold treatment is most practical when it can be adapted for use in commercial storage and shipping practices (Armstrong & Couey 1989). The treatment involves conditioning when required and ensuring that fruit is precooled to the target temperature and that the pulp temperatures are accurately monitored and controlled. The normal commercial practice of cold storage is adapted with minor modifications to become an effective disinfestation treatment. Cold storage disinfests commodities and extends their market life during transit.

Chilling injury can damage the fruit peel, pulp, and texture, delay or interrupt ripening, or enhance susceptibility to decay organisms (Morris 1982, Wang 1982, Harker & Hallet 1992). Many subtropical and tropical fruits cannot tolerate cold temperatures for the time required to provide quarantine security (Couey 1982). Some temperate zone warm season crops also develop chilling injury (Bramlage 1982).

Other disadvantages of cold storage are the cost of refrigeration to keep fruit cold for an extended duration of time and the amount of time needed for treatment. Monitoring of fruit pulp and air temperatures with accurate sensors and reliable recorders for extended times is required.

Fumigation or heat would probably be more economical than refrigeration (Clark & Weems 1989).

Potential for Cold Storage Quarantine Treatments

Preconditioning fruit may prevent chilling injury in some instances (Hatton & Cubbedge 1982, Houck et al. 1990, Wild & Hood 1989, Miller et al. 1990, Predebon & Edwards 1992). Other promising areas of research include growth regulators (Ismail & Grierson 1977, McDonald et al. 1988), other chemical treatments to make fruit less susceptible to chilling injury (Schiffmann-Nadel et al. 1972, Wardowski et al. 1975, Chalutz et al. 1985, McDonald et al. 1991), and plastic wrapping (Ben-Yehoshua 1985). Ripening the fruit to a less susceptible stage of ripeness has been used to reduce chilling injury (Grierson 1974, Chan 1988). Intermittent warming reduces chilling injury (Cohen et al. 1990) but may compromise quarantine treatments.

Conclusions

Cold storage is used as a quarantine treatment to ship fruit to areas where the fruit would otherwise not be permitted. Cold has a potential as a quarantine treatment especially when cold storage is used as part of the normal distribution and marketing procedure. The use of cold treatment may expand to other commodities as the problems of cold injury are resolved by research.

References

Anonymous. 1992a. Animal and Plant Health Inspection Service. Plant protection and quarantine treatment manual. T 107-Cold Treatment. U.S. Government Printing Office. Washington, D.C.

_____. 1992b. Animal and Plant Health Inspection Service. Plant protection and quarantine treatment manual. T 110-Quick Freeze. U.S. Government Printing Office. Washington, D.C.

Armstrong, J. W. & H. M. Couey. 1989. "Fumigation, Heat and Cold," in A. S. Robinson & G. Hooper, eds., World Crop Pests, Vol. 3B, Fruit Flies. Their Biology, Natural Enemies and Control. Pp. 411-424. Amsterdam: Elsevier.

Back, E. A. & C. E. Pemberton. 1916a. Effect of cold-storage temperatures upon the Mediterranean fruit fly. J. Agric. Research 5: 657-666.

_____. 1916b. Effect of cold temperatures upon the pupae of the Mediterranean fruit fly. J. Agric. Research 6: 251-260.

_____. 1918. The Mediterranean fruit fly in Hawaii. USDA Bul. 536. 118 pp.

Baker, A. C. 1952. "The Vapor-Heat Process," in *Insects: The Yearbook of Agriculture.* Pp. 401-404. Department of Agr. U.S. Government Printing Office. Washington, D.C.

Baker, A. C., W. E. Stone, C. C. Plummer & M. McPhail. 1944. A review of studies on the Mexican fruitfly and related Mexican species. USDA Miscellaneous Publication 531. 155 pp.

Baust, J. G. & R. R. Rojas. 1985. Review-Insect cold hardiness: Facts and fancy. J. Insect Physiology 31: 755-759.

Benschoter, C. A. 1979. Seasonal variation in tolerance of Florida 'Marsh' grapefruit to a combination of methyl bromide fumigation and cold storage. Proceedings, Fla. State Hortic. Soc. 92: 166-167.

_____. 1981. Tolerance of Florida 'Marsh' grapefruit to methyl bromide fumigation and cold storage combination treatments effective against the Caribbean fruit fly. Proceedings, Fla. State Hortic. Soc. 94: 301-303.

_____. 1983. Lethal effects of cold storage temperatures on Caribbean fruit fly in grapefruit. Proceedings, Fla. State Hortic. Soc. 96: 318-319.

_____. 1984. Low-temperature storage as a quarantine treatment for the Caribbean fruit fly (Diptera: Tephritidae) in Florida citrus. J. Econ. Entomol. 77: 1233-1235.

_____. 1987. Effects of modified atmospheres and refrigeration temperatures on survival of eggs and larvae of the Caribbean fruit fly (Diptera: Tephritidae) in laboratory diet. J. Econ. Entomol. 80: 1223-1225.

Benschoter, C. A. & P. C. Witherell. 1984. Lethal effects of suboptimal temperatures on immature stages of *Anastrepha suspensa.* Fla. Entomologist 67: 189-193.

Ben-Yehoshua, S. 1985. Individual seal-packaging of fruit and vegetables in plastic film - A new postharvest technique. HortScience 20: 32-37.

Bramlage, W. J. 1982. Chilling injury of crops of temperate origin. HortScience 17: 165-168.

Burditt, A. K., Jr., & J. W. Balock. 1985. Refrigeration as a quarantine treatment for fruits and vegetables infested with eggs and larvae of *Dacus dorsalis* and *Dacus cucurbitae* (Diptera: Tephritidae). J. Econ. Entomol. 78: 885-887.

Burditt, A. K., Jr., & L. C. McAlister, Jr. 1982. Refrigeration as a quarantine treatment for fruit infested with eggs and larvae of *Anastrepha* species. Proceedings, Fla. State Hortic. Soc. 95: 224-226.

Burikam, I., O. Sarnthoy, K. Charernsom, T. Kanno & H. Homma. 1992. Cold temperature treatment for mangosteens infested with the oriental fruit fly (Diptera: Tephritidae). J. Econ. Entomol. 85: 2298-2301.

Chalutz, E., J. Waks & M. Schiffmann-Nadel. 1985. Reducing susceptibility of grapefruit to chilling injury during cold treatment. HortScience 20: 226-228.

Chan, H. T., Jr. 1988. Alleviation of chilling injury in papayas. HortScience 23: 868-870.

Chen, C. P., R.E.J. Lee & D. L. Denlinger. 1990. A comparison of the responses of tropical and temperate flies (Diptera: Sarcophagidae) to cold and heat stress. J. of Comparative Physiology B. 160: 543-547.

Clark, R. & H. V. Jr. Weems. 1989. Detection, quarantine, and eradication of fruit flies invading Florida. Proceedings, Fla. State Hortic. Soc. 102: 159-164.

Clayton, K. C. 1993. Importation of fruits and vegetables. Federal Register 58: 11383-11389.

Cohen, E., S. Ben-yehoshua, I. Rosenberger, Y. Shalom & B. Shapiro Aro. 1990. Quality of lemons sealed in high-density polyethylene film during long-term storage at different temperatures with intermittent warming. J. Hortic. Sci. 65: 603-610.

Couey, H. M. 1982. Chilling injury of crops of tropical and subtropical origin. HortScience 17: 162-165.

Czajka, M. & R.E.J. Lee. 1990. A rapid cold-hardening response protecting against cold shock injury in *Drosophila melanogaster*. J. Experimental Biology 148: 245-254.

Davidson, J. K. 1990. Non-parallel geographic patterns for tolerance to cold and desiccation in *Drosophila melanogaster* and *D. simulans*. Australian J. Zool. 38: 155-161.

De Lima, C.P.F. 1992. Disinfestation of kiwifruit using cold storage as a quarantine treatment for Mediterranean fruit fly (*Ceratitis capitata* Wiedemann). New Zealand J. Crop & Hortic. Sci. 20: 223-227.

Dustan, G. G. 1963. The effect of standard cold storage and controlled atmosphere storage on survival of larvae of the oriental fruit moth, *Grapholita molesta*. J. Econ. Entomol. 56: 167-169.

Duvel, J.W.T. 1905. Cold storage for cowpeas. U.S. Department Agr. Bureau Entomol. Bul. 54: 49-54.

Enkerlin, D., L. Garcia R. & F. Lopez M. 1989. "Mexico, Central and South America," in A. S. Robinson & G. Hooper, eds., *World Crop Pests, Vol. 3A, Fruit Flies. Their Biology, Natural Enemies and Control*. Pp. 83-90. Amsterdam: Elsevier.

Fuller, C. 1906. Cold storage as a factor in the spread of insect pests. Natal Agric. J. and Mining Record 9: 656.

Gould, W. P. & J. L. Sharp. 1990. Cold-storage quarantine treatment for carambolas infested with the Caribbean fruit fly (Diptera: Tephritidae). J. Econ. Entomol. 83: 458-460.

Grierson, W. 1974. Chilling injury in tropical and subtropical fruit: Effect of harvest date, degreening, delayed storage and peel color on chilling injury to grapefruit. Proceedings, Tropical Region, American Soc. Hortic. Sci. 18: 66-73.

Harker, F. R. & I. C. Hallett. 1992. Physiological changes associated with development of mealiness of apple fruit during cool storage. HortScience 27: 1291-1294.

Hatton, T. T. & R. H. Cubbedge. 1982. Conditioning Florida grapefruit to reduce chilling injury during low-temperature storage. J. American Soc. Hortic. Sci. 107: 57-60.

Hill, A. R., C. J. Rigney & A. N. Sproul. 1988. Cold storage of oranges as a disinfestation treatment against the fruit flies *Dacus tryoni* (Froggatt) and *Ceratitis capitata* (Wiedemann) (Diptera: Tephritidae). J. Econ. Entomol. 81: 257-260.

Hooper, T. 1907. Cold storage and the fruit fly. J. of the Dept. of Agr., Western Australia 15: 252-253.

Houck, L. G., J. F. Jenner & B. E. Mackey. 1990. Seasonal variability of the response of desert lemons to rind injury and decay caused by quarantine cold treatments. J. Hortic. Sci. 65: 611-617.

Howard, L. O. 1896. Some temperature effects on household insects. U.S. Dept. Agr. Division Entomol. Bul. 6: 13-17.

Ismail, M. A. & W. Grierson. 1977. Seasonal susceptibility of grapefruit to chilling injury as modified by certain growth regulators. HortScience 12: 118-120.

Jessup, A. J. 1992. Low-temperature storage as a quarantine treatment for table grapes infested with Queensland fruit fly (*Bactrocera tryoni* Froggatt). New Zealand J. Crop & Hortic. Sci. 20: 235-239.

Jessup, A. J. & A. Baheer. 1990. Low-temperature storage as a quarantine treatment for kiwifruit infested with *Dacus tryoni* (Diptera: Tephritidae). J. Econ. Entomol. 83: 2317-2319.

Joint Legislative Committee on Agriculture and Livestock Problems. 1953. Third Special Report on the Control of the Oriental Fruit Fly (*Dacus dorsalis*) in the Hawaiian Islands. Senate of the State of California. 139 pp.

Lounsbury, C. P. 1907. The fruit fly (*Ceratitis capitata*). Agric. J. Cape Good Hope 31: 186-187.

Mason, A. C. & O. C. McBride. 1934. Effect of low temperature on the Mediterranean fruit fly in infested fruit. J. Econ. Entomol. 27: 897-902.

McDonald, R. E., P. D. Greany, P. E. Shaw, W. J. Schroeder, T. T. Hatton & C. W. Wilson. 1988. "Use of Gibberellic Acid for Caribbean Fruit

Fly (*Anastrepha suspensa*) Control in Grapefruit" in R. Goren & K. Mendel, eds., *Proceedings, 6th International Citrus Congress, Tel Aviv.* Pp. 37-43. Margraf Sci. Books: Weiker-sheim.

McDonald, R. E., W. R. Miller & T. G. McCollum. 1991. Thiabendazole and Imazalil applied at 53°C reduce chilling injury and decay of grapefruit. HortScience 26: 397-399.

Meats, A. 1976. Thresholds for cold-torpor and cold-survival in the Queensland fruit fly, and predictability of rates of change in survival threshold. Insect Physiology 22: 1505-1509.

Miller, W. R., D. Chun, L. A. Risse, T. T. Hatton & R. T. Hinsch. 1990. Conditioning of Florida grapefruit to reduce peel stress during low-temperature storage. HortScience 25: 209-211.

Moffitt, H. R. & D. J. Albano. 1972. Effects of commercial fruit storage on stages of the codling moth. J. Econ. Entomol. 65: 770-773.

Moffitt, H. R. & A. K. Burditt, Jr. 1989. Effects of low temperatures on three embryonic stages of the codling moth (Lepidoptera: Tortricidae). J. Econ. Entomol. 82: 1379-1381.

Morris, L. L. 1982. Chilling injury of horticultural crops: an overview. HortScience 17: 162-165.

Nel, R. G. 1936. The utilization of low temperatures in the sterilization of deciduous fruit infested with the immature stages of the Mediterranean fruit fly, *Ceratitis capitata* Wied. Union of South Africa Department Agr. and Forestry Sci. Bul. 155. 33 pp.

Pettey, F. W. & E. A. Griffiths. 1931. Effective control of fruit fly by refrigeration. Report on cold storage tests for the control of fruit fly, 1929-1930. Union of South Africa Department Agr. and Forestry Sci. Bul. 99. 9 pp.

Predebon, S. & M. Edwards. 1992. Curing to prevent chilling injury during cold disinfestation and to improve the external and internal quality of lemons. Australian J. of Experimental Agr. and Animal Husbandry 32: 233-236.

Richardson, H. H. 1952. "Cold treatment of fruits," in *USDA Yearbook of Agr.* Pp. 404-406.

_____. 1958. Treatments of various fruits and vegetables to permit their movement under fruit fly quarantines. Proceedings, Tenth International Congress of Entomol. (1956) 3: 17-23.

Ruckelshaus, W. D. 1984. Ethylene dibromide, amendment of notice of intent to cancel registration of pesticide products containing ethylene dibromide. Federal Register 49: 14182-14185.

Schiffmann-Nadel, M., E. Chalutz & F. S. Lattar. 1972. Reduction of pitting of grapefruit by thiabendazole during long term cold storage. HortScience 7: 394-395.

Sproul, A. N. 1976. Disinfestation of Western Australian Granny Smith apples by cold treatment against the egg and larval stages of the Mediterranean fruit fly [*Ceratitis capitata* (Wied.)]. Australian J. of Experimental Agr. and Animal Husbandry 16: 280-285.

Stout, O. O. (Revised by H. L. Roth). 1983. In J. F. Karpati, C. Y. Schotman & K. A. Zammarano, eds., *International Plant Quarantine Treatment Manual. FAO Plant Production and Protection Paper 50.* Pp. x+220. Rome, Italy: Food and Agr. Organization of the United Nations.

Union of South Africa, Department of Agriculture. 1923. Fruit-fly and cold storage. J. Dept. Agr., Union South Africa 7: 364-365.

Wang, C. Y. 1982. Physiological and biochemical responses of plants to chilling stress. HortScience 17: 162-165.

Wardowski, W. F., L. G. Albrigo, W. Grierson, C. R. Barmore & T. A. Wheaton. 1975. Chilling injury and decay of grapefruit as affected by thiabendazole, benomyl, and carbon dioxide. HortScience 10: 381-383.

Wilcox, E. V. & C. J. Hunn. 1914. Cold storage for tropical fruits. Hawaii Agric. Experiment Station Press Bul. 47: 1-12.

Wild, B. L. & C. W. Hood. 1989. Hot dip treatments reduce chilling injury in long-term storage of 'Valencia' oranges. HortScience 24: 109-110.

Yokoyama, V. Y. & G. T. Miller. 1989. Response of codling moth and oriental fruit moth (Lepidoptera: Tortricidae) immatures to low-temperature storage of stone fruits. J. Econ. Entomol. 82: 1152-1156.

9

Hot Water Immersion

Jennifer L. Sharp

Since the late 1800s heated water has been applied as a spray or an immersion treatment to fruits, vegetables, roots, bulbs, and stems. Heated water reduces the presence of pathogenic viruses, nematodes, fungi, bacteria, and populations of snail, mite, and insect pests (Koidsumi & Shibata 1936, Cohen & Nadel 1958, Anonymous 1972, Stout 1983, Roth 1989). Numerous hot water treatments are approved by the United States Department of Agriculture, Animal and Plant Health Inspection Service, Plant Protection and Quarantine (USDA-APHIS-PPQ). Treatments are published in the USDA-APHIS-PPQ Treatment Manual (Anonymous 1992).

In 1948, in the eastern United States (U.S.), an aqueous dip containing ethylene dibromide (EDB) combined with a residual toxin was used as a quarantine dip treatment to kill Japanese beetle, *Popillia japonica* Newman. Insects were present in balled or potted nursery plants (Burditt et al. 1963). Hot water was first used in fruit fly quarantine programs in the U.S. in 1953. Mexican fruit fly, *Anastrepha ludens* (Loew), immatures in mango and fruit fly infestations in papaya were killed when the fruits were dipped in an aqueous solution of EDB (Burditt et al. 1963). From 1970 to 1980, hot water dips followed by fumigation with EDB were used often as combination treatments. Hot water dips reduced postharvest fruit rots and killed Tephritidae eggs and larvae in tropical and subtropical fruits (Seo et al. 1972, Lin et al. 1976, Couey et al. 1985). After EDB was abolished as a commodity treatment for fruits, hot water dips were used alone primarily to disinfest papaya and mango of tephritid immatures to allow entry of treated fruits into the U.S.

Herein I discuss hot water as a postharvest-quarantine dip treatment used to provide quarantine security, or 99.9968% mortality (Baker 1939), against introduction and movement of various state and federal regulated pests of tropical and subtropical fruits.

Major Treatment Application

The USDA-APHIS-PPQ certified commercial hot-water treatment facilities for papaya grown in Hawaii and for mango grown in Mexico, Haiti, Puerto Rico, and South America. A typical commercial hot-water facility is shown in Figure 9.1. Hot-water immersion facilities have the capacity to treat thousands of fruits daily. All treatments must be in compliance with the USDA-APHIS-PPQ rules and regulations. Immersion in hot water has been used to disinfest banana, stone fruits, grapefruit, carambola, and guava of quarantine pests. The treatment is not recommended for grapefruit and stone fruits because it produces unacceptable damage at immersion times required to provide quarantine security.

Papaya

'Sunrise' and Kapoho 'Solo' papayas shipped to Japan and to the mainland U.S. from Hawaii are subjected to quarantine because of possible infestations by oriental fruit fly, *Bactrocera dorsalis* (Hendel), melon fly, *B. cucurbitae* (Coquillett), and Mediterranean fruit fly, *Ceratitis capitata* (Wiedemann) (Anonymous 1992). Also, papayas grown in Hawaii are immersed in hot water to kill surface anthracnose organisms (Akamine & Arisumi 1953).

In 1953 in Hawaii, water was mixed with EDB and tested as a dip treatment to kill fruit fly infestations in papaya (F. G. Hinman, unpublished data cited by Burditt et al. 1963). Burditt et al. (1963) refined the treatment by using 43.3 or 46.1°C water mixed with EDB. For several years, the standard treatment for papayas was 8 g per cubic meter of EDB for 2 h combined with immersion in 49°C water for 20 min. If the hot water treatment was omitted, the quantity of EDB was doubled.

Beginning 28 September 1983 the Environmental Protection Agency (EPA) began removing EDB from the market place. Action was needed because of evidence that its continued use posed risks of long term health effects which outweighed its benefit in protecting agricultural commodities from pests (Anonymous 1983). An emergency suspension was issued immediately that stopped its sale and distribution for soil fumigation use, cancelled its use as a fumigant for stored grain and grain milling machinery, and banned it for quarantine treatments of citrus and papaya. The suspension was scheduled to become effective on 1 September 1984 to allow time for the development of alternate treatment methods to prevent the spread of fruit fly infestations (Anonymous 1983).

Couey et al. (1984) proposed a systems approach as an alternate treatment for EDB. Treatment required that papayas were harvested between colorbreak and one-quarter ripe, immersed in 49°C water for 20 min, exposed to a 20°C water spray for 20 min, and refrigerated at 8 - 10°C for 10 d. Usually papayas one-quarter ripe or less do not have fruit fly infestations (Liquido et al. 1989). Benzyl isothiocyanate in papayas deters oviposition and increases egg and larval mortalities (Seo et al. 1982, 1983, Seo & Tang 1982).

On 2 March 1984, the EPA announced that it would establish a short term tolerance (legal residue level) of EDB for fruits, including papaya.

FIGURE 9.1. Commercial hot-water dip facility used to disinfest fruits of tephritid immatures.

The legal residue level was set at 250 parts per billion (ppb) of which no more than 30 ppb could be present in the edible pulp (Anonymous 1984). To comply with the 30 ppb tolerance level, Couey et al. (1985) omitted hydrocooling the papayas after a hot water treatment, fumigated the papayas at fruit pulp temperature of ≥33°C, and reduced the EDB dose. Fruits exceeding the established levels were not allowed in interstate commerce and subjected to enforcement action by the U.S.

Hayes et al. (1984) proposed using microwaves as an alternate to EDB. A microwave treatment of 2,450 Mhz was applied to papayas until the centers reached 38 - 45°C. Then the papayas were placed in 48.7°C water for 8 - 20 min and in 24°C water for 20 min. The combination treatment provided quarantine security; however, the combination treatment was never confirmed.

In 1987, Hayes et al. showed that the double immersion treatment did not provide quarantine security against oriental fruit fly in papayas one-half to three-quarters ripe. Seven live larvae were recovered from 15,540 treated eggs. Couey & Hayes (1986) reported that papayas less than one-quarter ripe immersed in 42°C water for 30 min (to preheat the papayas and to control papaya anthracnose decay organisms) and immediately transferred to 49°C water for 20 min (to kill eggs and larvae near the surface) produced unacceptable damage. The damage was characterized by hard areas in the papaya pulp that never ripened. When the interior was heated to lethal temperature, the fruit was damaged. Problems continued to be reported with the treatment. On 4 March 1987 and 12 July 1989 personnel of the California Department of Food and Agriculture in Oakland found viable third instar fruit flies in papayas that had been treated with the double immersion process (Zee et al. 1989). Preliminary observations of fruit returned to Hawaii by the USDA Agricultural Research Service (ARS) personnel indicated that "navel" and "pinhole" blossom end defects (abnormal placental growth near the blossom end) were responsible for the presence of living fruit fly larvae. The defects provided larval access into the papaya cavity (Liquido 1990). The double dip treatment interfered with the ripening process, did not kill all larvae present deep within the pulp in morphologically defective papaya, and was abolished by the USDA-APHIS-PPQ (Anonymous 1991).

Mango

Mangoes imported into the U.S. from Mexico, Central and South America, and Caribbean countries are subjected to quarantine because of possible infestation by Mediterranean fruit fly, many *Anastrepha* species, and mango weevil, *Cryptorhynchus mangiferae* (F.) (Anonymous 1992). Mangoes grown in Florida for shipment to Texas, Arizona, California, and Hawaii require treatment with approved state quarantine treatments because they serve as hosts for Caribbean fruit fly, *Anastrepha suspensa* (Loew) (Swanson & Baranowski 1972). Hawaii-grown mangoes cannot be shipped to mainland U.S. because no approved quarantine treatment kills fruit flies and mango weevil.

In 1971, research was initiated in Hawaii to determine if the hot water treatment used for postharvest control of anthracnose on mangoes (Pennock & Maldonaldo 1962, Smoot & Segall 1963) could be combined with low doses of EDB and refrigeration to provide quarantine security against fruit fly immatures in Hawaii-grown mangoes. Indeed, eggs and larvae of melon fly, oriental fruit fly, and Mediterranean fruit fly were killed in mangoes sequentially immersed in 46.3°C water for 120 min, fumigated with EDB in wooden field boxes and then refrigerated (Seo et al. 1972). A similar treatment was reported by Lin et al. (1976) who found no oriental fruit fly and melon fly larvae in Taiwan-grown mangoes immersed in 48 - 50°C water for 20 min, hydrocooled, dried and cooled, and then fumigated with EDB.

The first alternate treatment to EDB for mangoes was reported by Sharp & Spalding (1984) who showed that Caribbean fruit fly immatures were killed in Florida-grown mangoes immersed in 46.1 - 46.7°C water for 65 min without damaging the fruit. Although probit analysis estimated that quarantine security would be achieved against Caribbean fruit fly immatures in 'Tommy Atkins' and 'Keitt' mangoes treated for 68 min (Sharp 1986), the treatment was confirmed for 90 min to ensure quarantine security if other *Anastrepha* species or Mediterranean fruit fly invade Florida (Sharp et al. 1989a).

Research was initiated in 1986 in Port-au-Prince, Haiti, to develop a hot-water immersion quarantine treatment for Haiti-grown mangoes for export to the U.S. A treatment was approved in 1987 for 'Francis' mangoes infested with West Indian fruit fly, *Anastrepha obliqua* (Macquart), and Caribbean fruit fly (Sharp et al. 1988). Treatment procedures and facility requirements were published in the USDA-APHIS-PPQ Treatment Manual (Anonymous 1992). The successful research performed in Florida and Haiti led to mango research in Mexico (Sharp et al. 1989b, 1989c), Puerto Rico (Segarra-Carmona et al. 1990), Peru (Sharp & Picho-Martinez 1990), and Brazil (Nascimento et al. 1992). Data were analyzed with the PROBIT procedure (SAS Institute 1985). The 99.9968% mortality estimate and 95% fiducial limits (FL) were obtained using formulae in Finney (1971). Probit analysis estimated that 99.9968% mortality for seven fruit fly species ranged from 39.7 to 113.4 min (mean = 66.1 min) (Table 9.1). Data were approved for specific tephritids and mango cultivars in each country where work was performed. Thus, hot water immersion was approved as an acceptable treatment for mangoes from Mexico except Chiapas (Anonymous 1988); all states in Mexico, Central America north of and including Costa Rica, Puerto Rico and the Virgin Islands into or through Guam, Hawaii, and the continental U.S. (Anonymous 1990a); and for varieties of mangoes from all of South America, Central America, and the West Indies (Aruba, Bonaire, Curacao, Margarita, Tortuga, and Trinidad and Tobago)(Anonymous 1990b). Fruit flies of concern were all *Anastrepha* species and Mediterranean fruit fly. Hot water treatment was not approved for use on mangoes from countries known to have mango weevil (Barbados, Dominica, French Guiana, Guadaloupe, Martinique, and St. Lucia) (P. Witherell, personal communication), nor Suriname, which is known to have a species of *Bactrocera* (Anonymous 1990c, Coan 1991).

Minor Treatment Application

Banana

Banana grown in Hawaii is prohibited for shipment to the continental U.S. unless treated in compliance with the USDA-APHIS-PPQ Treatment Manual to kill fruit flies. Armstrong (1982) developed a 15 min, 50°C hot-water dip treatment for Hawaii-grown 'Brazilian' bananas against Mediterranean fruit fly, oriental fruit fly, and melon fly that did not damage the fruit. The treatment was confirmed at the 99.9968% mortality level for Mediterranean fruit fly and oriental fruit fly but not for melon fly because bananas are not hosts for melon fly in nature, and the fly is difficult to rear in banana.

Stone Fruits

Stone fruits (peaches, nectarines, apricots, cherries, plums) are hosts for fruit flies and moths. Oriental fruit moth, *Grapholita molesta* (Busck), is a pest of California stone fruits and a quarantine pest of nectarines shipped from the San Joaquin Valley to British Columbia, where the pest is not found (Monro 1958). Use of hot water to control oriental fruit moth in stone fruits is limited by the tolerance of the fruit to heat injury. Maximum temperature that stone fruits can tolerate is about 43°C, whereas 45°C for 40 min is needed to reach 100% mortality (Yokoyama & Miller 1987). Hot water immersion is not recommended as a quarantine treatment against Caribbean fruit fly immatures in stone fruits (Sharp 1990).

TABLE 9.1. Estimated probit 9 mortality and 95% fiducial limits (FL) for fruit fly immatures in mango cultivars

Probit 9 (min)	95% FL (min)	Species[a]	Origin and Strain	Mango Cultivar
66.8	59.5 - 78.1	*A. obliqua*	Haiti, lab	Kent
83.6	74.0 - 98.3	*A. obliqua*	Mex., wild	Kent
72.1	63.7 - 84.5	*A. obliqua*	Mex., wild	Kent
58.0	47.4 - 84.9	*A. obliqua*	Haiti, lab	Francis
58.7	46.8 - 88.2	*A. obliqua*	Puerto Rico, wild	Keitt
65.7	53.4 - 97.9	*A. obliqua* (egg)	Brazil, wild	Haden, Keitt, Tommy Atkins
65.2	54.3 - 87.7	*A. obliqua* (1-d-lar.)	Brazil, wild	Haden, Keitt, Tommy Atkins
113.4	93.2 - 150.2	*A. obliqua*	Peru, wild	Haden
71.4	68.7 - 74.5	*A. ludens*	Mex., wild	Kent, Haden
65.1	57.2 - 77.3	*A. ludens*	Tex., lab	Haden, Kent, Keitt, Tommy Atkins
70.8	62.8 - 83.0	*A. ludens*	Tex., lab	Keitt
75.7	61.7 - 105.3	*A. ludens*	Tex., lab	Kent
53.9	45.8 - 69.4	*A. ludens*	Tex., lab	Tommy Atkins
50.3	42.5 - 64.3	*A. ludens*	Tex., lab	Haden
56.0	43.3 - 91.1	*A. ludens*	Tex., lab	Francis
39.7	29.4 - 52.4	*A. fraterculus* (egg)	Brazil, wild	Haden, Keitt, Tommy Atkins
68.5	49.4 - 159.6	*A. fraterculus* (1-d-lar.)	Brazil, wild	Haden, Keitt, Tommy Atkins
75.6	70.1 - 82.5	*A. fraterculus*	Peru, wild	Haden
65.8	59.5 - 74.4	*A. distincta*	Peru, wild	Haden
76.1	66.2 - 91.4	*C. capitata*	Peru, wild, lab	Haden, Kent
59.4	49.6 - 83.8	*C. capitata* (egg)	Brazil, lab	Haden, Keitt, Tommy Atkins
79.7	70.4 - 93.7	*C. capitata* (l-d-lar.)	Brazil, lab	Haden, Keitt, Tommy Atkins
67.5	60.4 - 78.5	*C. capitata*	Mex., lab	Kent
44.3	39.1 - 53.3	*A. suspensa*	Fla., lab	Francis
60.5	58.9 - 62.3	*A. suspensa*	Fla., lab	Keitt
60.0	58.1 - 62.1	*A. suspensa*	Fla., lab	Tommy Atkins
62.0	60.6 - 63.5	*A. suspensa*	Mex., wild	Kent

[a] Estimates derived with mature, third instars in fruit unless other stage noted.

Grapefruit

Grapefruit grown in Florida is susceptible to infestation by Caribbean fruit fly and must be treated to prevent spread of the fly. Approved postharvest treatments for disinfesting Florida grapefruit of Caribbean fruit fly are cold temperature storage (Benschoter 1984) and methyl bromide fumigation (Benschoter 1979). These treatments occasionally produce damage and alternate methods are needed. Preliminary work by J. L. Sharp showed that Caribbean fruit fly immatures were killed in grapefruit immersed in 43.3°C water for 4 h. However, Miller et al. (1988) showed that grapefruit was damaged when immersed in 43.5°C for 4 h. Sharp (1985) used 48.9 - 57.2°C water for 10 - 40 min against Caribbean fruit fly eggs and larvae in Florida-grown grapefruit and reported that none of the time-temperature combinations produced quarantine security. Also, noninfested grapefruit immersed in 48.9°C water for 20 min exhibited severe scalding and pitting of the epidermis and produced off-flavors compared with grapefruit immersed in 26.7°C water for 40 min.

Gould (1988) tried hot water immersion and refrigeration to determine the potential of the combined method as a quarantine treatment. Probit analysis of data estimated that immersion in 43.3°C water for 100 min followed by 1.1°C storage for 7 d would provide quarantine security; however, additional tests must be done to confirm at least 100,000 larvae are killed with no survivors before the combined method could be approved by the USDA-APHIS-PPQ as a quarantine treatment.

Grapefruit is a host for eggs of Fuller rose beetle, *Asynonychus godmani* Crotch. Beetles are parthenogenic and oviposit beneath the calyxes of citrus. Japan will not accept citrus from the U.S. if viable eggs are found. Soderstrom & Brandl (1988) found that egg development was stopped by immersion of grapefruit in 55°C water for 2 min.

Carambola

Carambola is a tropical tree fruit originating in southeastern Asia. Carambola was originally introduced into Florida before 1887 but has recently become a commercially important fruit crop (Knight 1964, Campbell et al. 1985). Carambola serves as a host for Caribbean fruit fly (Swanson & Baranowski 1972) and must be treated with an approved quarantine treatment before shipment to Texas, Arizona, California, and Hawaii. Immersing carambolas in 46 - 46.4°C water for 45 min provided quarantine security against the fly but reduced the shelf life of treated fruit (Hallman & Sharp 1990a, Hallman 1991).

Guava

Guava is a host for Caribbean fruit fly in Florida (Swanson & Baranowski 1972). Gould & Sharp (1992) immersed guavas infested with mature Caribbean fruit fly larvae in 46.1°C water for 35 min. They reported that the treatment provided quarantine security and did not damage fruit quality.

Miscellaneous Commodities

The USDA-APHIS-PPQ prohibits the import into the U.S. of products that could serve as hosts for melon thrips, *Thrips palmi* Karny, unless the hosts have been treated and precleared. For example, each shipment of bittermelon, luffa, long bean, and eggplant may be imported from the Dominican Republic through all ports, subject to verification that the vegetables were dipped in 49°C water for 7 min and accompanied with a USDA-APHIS-PPQ Form 203 and a phytosanitary certificate affirming the treatment.

Conclusions

The USDA-ARS has been the leader in the development of hot water dips to disinfest commodities of pests of quarantine importance. Hot water dips for quarantine consist of immersing fruits in 43 - 46°C water. Temperatures <43°C are not practical in regard to current industry demands and temperatures >46°C tend to produce excessive commodity damage. Time that the fruits remain immersed in water to kill infestations without unacceptable damage to the treated commodity depends on the water temperature and circulation, fruit size and maturity, and the physical characteristics of the fruit such as thickness of the pulp. When mangoes are handled roughly and then treated by immersion in hot water, dark areas appear on the peel. The areas seem to be water soaked. Also, physiologically unripe fruits immersed in 46.1°C water will not ripen properly throughout the pulp which remains hard. Freshly picked mature green mangoes immersed in 46.1°C water occasionally will display some hard areas in the pulp. Preconditioning the mangoes for 1 - 2 d at room temperatures may allow for some additional fruit maturation and eliminate heat damage in some cases. Hydrocooling mangoes after treatment with hot water was suggested by industry to reduce heat damage to fruit quality. Research must demonstrate, however, that hydrocooling does not adversely affect quality and jeopardize quarantine security. Hallman & Sharp (1990b) reported that hydrocooling mangoes that had been treated with 46.1 - 46.7°C water for 54 min increased survival of mature Caribbean fruit fly larvae. The study provided evidence that hydrocooling mangoes immediately after hot water treatment could effect quarantine security. The quality of hot-water treated 'Keitt' mangoes was reduced by hydrocooling (R. McGuire & J. Sharp, unpublished data).

Hot water immersion continues to be a promising technology to disinfest commodities, especially mangoes that tolerate heat. Advantages of the treatment include relative ease of use by industry, short treatment time, reliable and accurate monitoring and certification by the USDA-APHIS-PPQ. The biorationale treatment leaves no toxic chemical residue, increases the market quality by killing surface decay organisms, and cleans the fruit of plant exudates. Use of hot water as a quarantine treatment could increase as chemical fumigants such as methyl bromide are discontinued. Detrimental effects to fruit quality by hot water should be reduced as more data are developed and as user groups become familiar with the treatment application. Improvements in predictive modeling of insect mortality and temperature-mortality relationships (Jang 1986, 1991, Hayes et al. 1987, Sharp & Chew 1987, Hansen et al. 1990, Heard et al. 1991), thermal diffusivity (Hayes 1984, Rodriguez & Gaffney 1988, Rodriguez et al. 1989), and studies of heat tolerance of commodities will reduce damage to commodities, improve their quality, and maintain needed quarantine security against pests.

Acknowledgments

I thank Albert P. Whitworth, Agri Machinery, Inc., Orlando, Florida, for providing the photograph used to illustrate a typical commercial hot water immersion facility. I also thank Sharon Pickard, USDA-ARS, for typing the manuscript.

References

Akamine, E. K. & T. Arisumi. 1953. Control of postharvest storage decay of fruits of papaya (*Carica papaya* L.) with special reference to the effect of hot water. Proceedings, American Soc. Hortic. Sci. 61: 270-274.

Anonymous. 1972. Hot-water treatment of plant material. Ministry of Agr., Fisheries and Food Bul. 201. London, 43 pp.

_____. 1983. Environmental Protection Agency. Ethylene dibromide; intent to cancel registrations of pesticide products containing ethylene dibromide; determination concluding the rebuttable presumption against registration; availability of position document. Federal Register 48: 46234-46248.

_____. 1984. Environmental Protection Agency. Rules and regulations. Revocation of tolerance for ethylene dibromide. Federal Register 49: 22082-22085.

_____. 1988. Rules and regulations. Plant protection and quarantine treatment manual; incorporation by reference. Federal Register 53: 10525-10526.

_____. 1990a. Hot water dip treatments for mangoes. Federal Register 55: 5433-5436.

_____. 1990b. Hot water dip treatments for mangoes. Federal Register 55: 39132-39134.

_____. 1990c. Outbreaks and new records. Suriname. Update on fruit-fly (*Dacus dorsalis*) situation. FAO Plant Protection Bul. 38:51.

_____. 1991. Department of Agriculture. Animal and Plant Health Inspection Service. Final Rule. Papayas from Hawaii. Federal Register 56: 59205-59207.

_____. 1992. Animal and Plant Health Inspection Service. Plant protection and quarantine treatment manual. U.S. Government Printing Office. Washington, D.C.

Armstrong, J. W. 1982. Development of a hot-water immersion quarantine treatment for Hawaiian-grown 'Brazilian' bananas. J. Econ. Entomol. 75: 787-790.

Baker, A.C. 1939. The basis for treatment of products where fruitflies are involved as a condition for entry into the United States. USDA Circular 551.

Benschoter, C. A. 1979. Fumigation of grapefruit with methyl bromide for control of *Anastrepha suspensa*. J. Econ. Entomol. 72: 401-402.

_____. 1984. Low temperature storage as a quarantine treatment for the Caribbean fruit fly (Diptera: Tephritidae) in Florida citrus. J. Econ. Entomol. 77: 1233-1235.

Burditt, A. K., Jr., J. W. Balock, F. G. Hinman & S. T. Seo. 1963. Ethylene dibromide water dips for destroying fruit fly infestations of quarantine significance in papayas. J. Econ. Entomol. 56: 289-292.

Campbell, C. W., R. J. Knight, Jr., & R. Olszack. 1985. Carambola production in Florida. Proceedings, Fla. State Hortic. Soc. 98: 145-149.

Coan, R. M. 1991. Nomenclature-carambola fruit fly. United States Department of Agriculture. Animal and Plant Health Inspection Service. International Services. May 31 memorandum to file. Hyattsville, MD.

Cohen, I. & D. Nadel. 1958. Instructions for using ethylene dibromide tank-dip method to control eggs and larvae of the Mediterranean fruit fly (*Ceratitis capitata*) in citrus fruit. (Trans. from Hebrew). Citrus Marketing Board of Israel, Technical Division. Pp. 1-16.

Couey, H. M. & C. F. Hayes. 1986. Quarantine procedure for Hawaiian papaya using fruit selection and a two-stage hot-water immersion. J. Econ. Entomol. 79: 1307-1314.

Couey, H. M., E. S. Linse & A. N. Nakamura. 1984. Quarantine procedure for Hawaiian papayas using heat and cold treatments. J. Econ. Entomol. 77: 984-988.

Couey, H. M., J. W. Armstrong, J. W. Hylin, W. Thornburg, A. N. Nakamura, E. S. Linse, J. Ogata & R. Vetro. 1985. Quarantine procedure for Hawaii papaya using a hot-water treatment and high-temperature, low-dose ethylene dibromide fumigation. J. Econ. Entomol. 78: 879-884.

Finney, D. J. 1971. *Probit Analysis, 3rd Ed.* Cambridge University Press.

Gould, W. P. 1988. A hot water/cold storage quarantine treatment for grapefruit infested with the Caribbean fruit fly. Proceedings, Fla. State Hortic. Soc. 101: 190-192.

Gould, W. P. & J. L. Sharp. 1992. Hot-water immersion quarantine treatment for guavas infested with Caribbean fruit fly (Diptera: Tephritidae). J. Econ. Entomol. 85: 1235-1239.

Hallman, G. J. 1991. Quality of carambolas subjected to postharvest hot water immersion and vapor heat treatments. HortScience 26: 286-287.

Hallman, G. J. & J. L. Sharp. 1990a. Hot-water immersion quarantine treatment for carambolas infested with Caribbean fruit fly (Diptera: Tephritidae). J. Econ. Entomol. 83: 1471-1474.

_____. 1990b. Mortality of Caribbean fruit fly (Diptera: Tephritidae) larvae infesting mangoes subjected to hot-water treatment, then immersion cooling. J. Econ. Entomol. 83: 2320-2323.

Hansen, J. D., J. W. Armstrong, B.K.S. Hu & S. A. Brown. 1990. Thermal death of oriental fruit fly (Diptera: Tephritidae) third instars in developing quarantine treatments for papayas. J. Econ. Entomol. 83: 160-167.

Hayes, C. F. 1984. Thermal diffusivity of papaya fruit (*Carica papaya* L. var. Solo). J. Food Sci. 49: 1219, 1221.

Hayes, C. F., H.T.G. Chingon, F. A. Nitta & W. J. Jang. 1984. Temperature control as an alternative to ethylene dibromide fumigation for the control of fruit flies (Diptera: Tephritidae) in papaya. J. Econ. Entomol. 77: 683-686.

Hayes, C. F., H.T.G. Chingon, F. A. Nitta & A.M.T. Leung. 1987. Calculation of survival from double hot-water immersion treatment for papayas infested with oriental fruit flies (Diptera: Tephritidae). J. Econ. Entomol. 80: 887-890.

Heard, T. A., N. W. Heather & R. J. Corcoran. 1991. Dose-mortality relationships for eggs and larvae of *Bactrocera tryoni* (Diptera: Tephritidae) immersed in hot water. J. Econ. Entomol. 84: 1768-1770.

Jang, E. 1986. Kinetics of thermal death in eggs and first instars of three species of fruit flies (Diptera: Tephritidae). J. Econ. Entomol. 79: 700-705.

_____. 1991. Thermal death kinetics and heat tolerance in early and late third instars of the oriental fruit fly (Diptera: Tephritidae). J. Econ. Entomol. 84: 1298-1303.

Knight, R. J., Jr. 1964. The carambola in south Florida. Rare Fruit Council of South Florida, Incorporated 1: 3-5, 12.

Koidsumi, K. & K. Shibata. 1936. On the velocity of heating and cooling of some fruits (Information for the control of fruit flies). J. Soc. Trop. Agr. (Japan) 8: 82-94 (in Japanese).

Lin, T. H., F. C. Tseng, C. R. Chang & L. Y. Wang. 1976. Multiple treatment for disinfesting oriental fruit fly in mangoes. Plant Protection Bul. 18: 231-241.

Liquido, N. J. 1990. Survival of oriental fruit fly and melon fly (Diptera: Tephritidae) eggs oviposited in morphologically defective blossom end of papaya following two-stage hot-water immersion treatment. J. Econ. Entomol. 83: 2327-2330.

Liquido, N. J., R. T. Cunningham & H. M. Couey. 1989. Infestation rates of papaya by fruit flies (Diptera: Tephritidae) in relation to the degree of fruit ripeness. J. Econ. Entomol. 82: 213-219.

Miller, W. R., R. E. McDonald, T. T. Hatton & M. Ismail. 1988. Phytotoxicity to grapefruit exposed to hot water immersion treatment. Proceedings, Fla. State Hortic. Soc. 101: 192-195.

Monro, H.A.U. 1958. Eradication measures against the oriental fruit moth in the province of British Columbia, Canada. FAO Plant Protection Bul. 6: 177-179.

Nascimento, A. S., A. Malavasi, J. S. Morgante & A.L.A. Duarte. 1992. Hot-water immersion treatment for mangoes infested with *Anastrepha fraterculus*, *A. obliqua*, and *Ceratitis capitata* (Diptera: Tephritidae) in Brazil. J. Econ. Entomol. 85: 456-460.

Pennock, W. & G. Maldonaldo. 1962. Hot-water treatment of mango fruits to reduce anthracnose decay. J. Agr. Univ. Puerto Rico 46: 272-283.

Rodriguez, A. C. & J. J. Gaffney. 1988. Properties of mangos as related to heat and mass transfer. American Soc. Agric. Engineers. Paper 88-6580.

Rodriguez, A. C., G. J. Hallman, W. P. Gould & J. J. Gaffney. 1989. Modelling fruit quarantine heat treatments. American Soc. Agric. Engineers. Paper 89-6053.

Roth, H. 1989. Concepts and recent developments in regulatory treatments, in R. P. Kahn, ed., *Plant Protection and Quarantine, Vol. 3.*

Pp. 117-144. Boca Raton, Florida: CRC Press.

SAS Institute, Inc. 1985. SAS user's guide: statistics. SAS Institute, Cary, North Carolina.

Segarra-Carmona, A. E., R. A. Franqui, L. V. Ramirez-Ramos, L. R. Santiago & C. N. Torres-Rivera. 1990. Hot water dip treatments to destroy *Anastrepha obliqua* larvae (Diptera: Tephritidae) in mangoes from Puerto Rico. J. Agr. University of Puerto Rico 74: 441-447.

Seo, S. T. & C. Tang. 1982. Hawaiian fruit flies (Diptera: Tephritidae): Toxicity of benzyl isothiocyanate against eggs of first instars of three species. J. Econ. Entomol. 75: 1132-1135.

Seo, S. T., D. L. Chambers, E. K. Akamine, M. Komura & C.Y.L. Lee. 1972. Hot water-ethylene dibromide fumigation-refrigeration treatment for mangoes infested by oriental and Mediterranean fruit flies. J. Econ. Entomol. 65: 1372-1374.

Seo, S. T., G. J. Farias & E. J. Harris. 1982. Oriental fruit fly: ripening of fruit and its effect on the index of infestation of Hawaiian papayas. J. Econ. Entomol. 75: 173-178.

Seo, S. T., C. S. Tang, S. Sanidad & T. H. Takenaka. 1983. Hawaiian fruit flies (Diptera: Tephritidae): variation of index of infestation with benzyl isothiocyanate concentration and color of maturing papayas. J. Econ. Entomol. 76: 535-538.

Sharp, J. L. 1985. Submersion of Florida grapefruit in heated water to kill stages of Caribbean fruit fly, *Anastrepha suspensa*. Proceedings, Fla. State Hortic. Soc. 98: 78-80.

_____. 1986. Hot-water treatment for control of *Anastrepha suspensa* (Diptera: Tephritidae) in mangos. J. Econ. Entomol. 79: 706-708.

_____. 1990. Immersion in heated water as a quarantine treatment for California stone fruits infested with the Caribbean fruit fly (Diptera: Tephritidae). J. Econ. Entomol. 83: 1468-1470.

Sharp, J. L. & V. Chew. 1987. Time/mortality relationships for *Anastrepha suspensa* (Diptera: Tephritidae) eggs and larvae submerged in hot water. J. Econ. Entomol. 80: 646-649.

Sharp, J. L. & H. Picho-Martinez. 1990. Hot-water quarantine treatment to control fruit flies in mangoes imported into the United States from Peru. J. Econ. Entomol. 83: 1940-1943.

Sharp, J. L. & D. H. Spalding. 1984. Hot water as a quarantine treatment for Florida mangos infested with Caribbean fruit fly. Proceedings, Fla. State Hortic. Soc. 97: 355-357.

Sharp, J. L., M. T. Ouye, R. Thalman, W. Hart, S. Ingle & V. Chew. 1988. Submersion of 'Francis' mango in hot water as a quarantine treatment for the West Indian fruit fly and the Caribbean fruit fly (Diptera: Tephritidae). J. Econ. Entomol. 81: 1431-1436.

Sharp, J. L., M. T. Ouye, W. Hart, S. Ingle, G. Hallman, W. Gould & V.

Chew. 1989a. Immersion of Florida mangos in hot water as a quarantine treatment for Caribbean fruit fly (Diptera: Tephritidae). J. Econ. Entomol. 82: 186-188.

Sharp, J. L., M. T. Ouye, S. J. Ingle & W. G. Hart. 1989b. Hot-water quarantine treatment for mangoes from Mexico infested with Mexican fruit fly and West Indian fruit fly (Diptera: Tephritidae). J. Econ. Entomol. 82: 1657-1662.

Sharp, J. L., M. T. Ouye, S. J. Ingle, W. G. Hart, W. R. Enkerlin H., H. Celedonio H., J. Toledo A., L. Stevens, E. Quintero, J. Reyes F. & A. Schwarz. 1989c. Hot-water quarantine treatment for mangoes from the state of Chiapas, Mexico, infested with Mediterranean fruit fly and *Anastrepha serpentina* (Wiedemann) (Diptera: Tephritidae). J. Econ. Entomol. 82: 1663-1666.

Smoot, J. J. & R. H. Segall. 1963. Hot water as a postharvest control of mango anthracnose. Plant Disease Reporter 47: 739-742.

Soderstrom, E. L. & D. G. Brandl. 1988. Hot-water dip for control of Fuller rose beetle eggs, 1986. Insecticide and Acaricide Tests 13: 360.

Stout, O. O. (Revised by H. L. Roth). 1983. In J. F. Karpati, C. Y. Schotman & K. A. Zammarano, eds., *International Plant Quarantine Treatment Manual. FAO Plant Production and Protection Paper 50.* Pp. x+220. Rome, Italy: Food and Agr. Organization of the United Nations.

Swanson, R. W. & R. M. Baranowski. 1972. Host range and infestation by the Caribbean fruit fly, *Anastrepha suspensa* (Diptera: Tephritidae), in south Florida. Proceedings, Fla. State Hortic. Soc. 84: 271-274.

Yokoyama, V. & G. T. Miller. 1987. High temperature for control of oriental fruit moth (Lepidoptera: Tortricidae) in stone fruits. J. Econ. Entomol. 80: 641-645.

Zee, F. T., M. S. Nishina, H. T. Chan & K. A. Nishijima. 1989. Blossom end defects and fruit fly infestation in papayas following hot water quarantine treatment. HortScience 24: 323-325.

10

Heated Air Treatments

Guy J. Hallman and John W. Armstrong

Quarantine treatments that subject fresh fruits and vegetables to heated air are vapor heat, referred to in some older literature as moist heat or heat sterilization, and high temperature forced air, commonly called forced hot air. Both treatments expose fresh commodities to heated air with water vapor added to increase the heat holding capacity of the air and to prevent the commodity from desiccating. The difference between these two treatments is that vapor heat uses saturated or nearly saturated air (100% relative humidity or RH) and transfers heat to the fruit surface through the condensation of water vapor and convection from heated air. In a forced hot air treatment RH may be as low as 30% and may fluctuate during treatment; heat transfer is by convection only. No condensation should form on the fruit surface during a forced hot air treatment.

Other heated air treatments used in quarantine are dry heat and steam sterilization (FAO 1983, Anonymous 1985). Dry heat uses air at usually 80 - 100°C without added humidity and is applied to soil, dry animal feed, and milled products. One variation of dry heat is the treatment of sweet potatoes at 39.4°C for 30 h to kill root knot nematodes, *Meloidogyne* species. Steam sterilization uses saturated air at 100 - 120°C, sometimes under pressure, to control pests and disease spores in straw and similar nonfood items. Dry heat and steam sterilization are generally not used for fresh fruits and vegetables, because they are too damaging, and will not be covered here. The term dry heat should not be used interchangeably with forced hot air (Cowley et al. 1992) as they are two distinct treatments. Likewise, the term hot dry air should not be used for forced hot air (Waddell & Birtles 1992).

In this chapter we discuss the history, research, and present use of vapor heat and hot air treatments.

Vapor Heat

Crawford (1927, reprinted as Crawford 1929) published the first account of using heated air to kill fruit fly larvae inside fruits. Few details were given; Mexican fruit fly, *Anastrepha ludens* (Loew), larvae were killed in unspecified fruit heated to 46°C for 24 h in a banana ripening room. Larvae were killed also by placing infested fruits in a "steam heater" at 43.3°C for a few hours.

The first commercial development of heated air as a quarantine treatment was vapor heat quarantine treatment for Mediterranean fruit fly, *Ceratitis capitata* (Wiedemann), in Florida in 1929 (Hawkins 1930, Latta 1932, Baker 1952). Commercial vapor heat treatment facilities were modified citrus degreening rooms or rooms built specifically for vapor heat treatment (Hawkins 1932). Steam was mixed with air and a fine water spray to reach 43.3°C. Under these conditions the RH stayed near 100%. The saturated heated air was forced into the treatment chamber (7.1 by 4.9 by 2.5 meters high) which held one railroad car load of citrus packed in 400 field boxes. The treatment consisted of gradually warming infested fruits for several hours (approach time) until the fruit interior reached 43.3°C and then holding that temperature for 8 h. The approach time was recommended to be between 6 - 8 h to avoid damaging the fruit (Hawkins 1929). High humidity was necessary to prevent damage to citrus (Jones et al. 1939). An air flow rate of 146 - 175 cubic meters per min moved the heated air over the fruit load. The air exiting the fruit load was recycled through the steam system where it was reheated and replenished with water vapor.

More than 81,000 metric tons of citrus were treated with vapor heat in Florida before Mediterranean fruit fly was eradicated (Jones et al. 1939).

Following the initial commercial application of vapor heat in Florida, vapor heat research was begun in other geographical areas. Mackie (1931) studied the effect of vapor heat at 43.3°C for 14 h on many different fruits and vegetables in California in the event that Mediterranean fruit fly was found in that state. His favorable early results on deciduous fruits and avocados were later found to be overly optimistic (Claypool & Vines 1956, Sinclair & Lindgren 1955).

Weddell (1931) tested vapor heat on the ability to kill eggs and larvae of Queensland fruit fly, *Bactrocera tryoni* (Froggatt), in several fruits, discovering that citrus tolerated the treatment whereas apples did not.

Benlloch (1934) constructed a small vapor heat treatment chamber in Mexico, and experimented with grapes and oranges infested with Mediterranean fruit fly. Grapes were treated inside 9.5-liter shipping crates. The entire treatment required 23 h; 15 h was required to heat the center of the crate to 44.8°C.

Seín (1935) investigated vapor heat treatments of mangoes and guavas infested with *Anastrepha* species (probably West Indian fruit fly, *A. obliqua* [Macquart], and Caribbean fruit fly, *A. suspensa* [Loew]). He was the first to attempt to shorten the time required for vapor heat treatments when he treated fruit fly infested mangoes and guavas for 4 h at 43°C instead of the standard 8 h and achieved 100% mortality of fruit fly immatures. He also killed 100% of fruit fly larvae in mangoes wrapped and crated for shipment (10 mangoes per crate) using 8 h of treatment at 43°C.

In 1932 the vapor heat treatment was approved for Texas grapefruit infested with Mexican fruit fly. From 1938, when it was first used, through the early 1950s, 508,000 metric tons of Texas grapefruit were treated with vapor heat. In 1945 vapor heat was approved for treating Mexican citrus and mangoes infested with Mexican fruit fly. The treatment allowed for the importation of these fruits into the United States (U. S.). Balock & Starr (1945) used probit analysis to estimate the amount of time necessary to achieve probit 9 (99.9968%) mortality in mangoes and found that disinfestation required a total treatment time of 13.7 h (which included an approach period of 8 h) at 43.3°C.

Vapor heat was studied for quarantine control of melon fly, *Bactrocera cucurbitae* (Coquillett), and oriental fruit fly, *Bactrocera dorsalis* (Hendel), in Taiwan and Okinawa. Koidsumi & Shibata (1936) found that the humidity content of the air was positively correlated to the speed with which a fruit heated. They developed equations to predict the time required to heat fruits. Koidsumi (1936a) studied the mortality of oriental fruit fly and melon fly eggs, larvae, and pupae in test tubes containing slices of citrus and cucumber exposed to vapor heat at temperatures between 40 - 46°C. Differences in the susceptibility of the various stages or species to heat were not apparent. Third instars of an unnamed species of fruit fly, presumably oriental fruit fly, were easier to kill in pummelo with vapor heat than eggs and first and second instars (Koidsumi 1936b). An earlier study found no apparent differences in susceptibility of oriental fruit fly eggs, all three instars, and pupae in test tubes with slices of citrus exposed to vapor heat at temperatures between 40 - 46°C (Koidsumi 1936a). Koidsumi (1936a) postulated that the deteriorated condition of the fruits containing third instars might have caused these fruits to heat faster than fruits containing other stages. However, temperatures inside the fruits were not recorded.

In 1938 work was initiated on vapor heat treatment of Hawaii-grown papayas infested with Mediterranean fruit fly and melon fly. The treatment consisted of an 8 h approach period at 43.3°C, followed by an 8 h holding time at that temperature. After some initial successes, the papayas began to show damage (Jones et al. 1939). It was hypothesized that fruits with high moisture content, such as those picked after a rainy period, were damaged by the vapor heat treatment and that if the water content of the fruits was reduced before treatment injury might be limited. Vapor heat treatments were designed that used a "conditioning period" of 6 h at 38°C, 60% RH, followed by an approach period from 2 h to 2 h and 30 min at 43.3°C, 60 - 80% RH, and the "sterilization period" of 8 h at 43.3°C, 100% RH. This was compared with a normal vapor heat treatment with the RH constant at 100%. Resulting damage to mature, green papayas was 100% for the vapor heat treatment at 100% RH and 0% for the vapor heat treatment with periods of 60 - 80% RH (Jones et al. 1939). Although differences in respiration rates of ripe papayas held at 43.3°C and 60% versus 100% RH for 24 h were found, it was unclear how these differences would affect the ability of papayas to tolerate vapor heat (Jones 1939). Jones (1940) recommended that papayas be treated with a conditioning period of 8 h at 38 - 43.3°C and 55 - 60% RH followed by 8 h at 43.3°C and 100% RH.

Balock & Kozuma (1954) researched higher treatment temperatures and shorter treatment times for papayas. Holding times as short as 1 h and 55 min at 47.2°C and 100% RH following an 8 h conditioning period at 47 - 48°C and 50% RH achieved 100% kill of oriental fruit fly immatures in papayas.

Akamine (1966) developed a vapor heat treatment for papayas which consisted of a preconditioning period at 43.3°C, 35% RH, for 6 h followed by 4 h of holding at 48°C and 100% RH. To reduce heat damage to the fruit, which appeared inconsistently throughout the harvest season, Seo et al. (1974) modified the treatment. They developed an 11 h approach period to raise the temperature of the fruit from 23.3 to 44.4°C, at which point the fruit were held at 44.4°C for an additional 8 h and 45 min. The treatment produced 99.999% mortality against oriental fruit fly.

Hawaiian fruits and vegetables other than papaya were studied for tolerance to vapor heat, both at 100% RH throughout the treatment and with preconditioning periods of <100% RH (Jones 1940). Precise data were not given; however, recommendations for vapor heat treatment of eggplant, lychee, chinese peas, green snap beans, yellow wax beans, lima beans, cucumbers, tomatoes, and bell peppers were provided. Avocados did not tolerate vapor heat.

The discovery of oriental fruit fly in Hawaii in 1946 stimulated further research in California on the effect of vapor heat on fruit quality in the event of an oriental fruit fly infestation in California (Sinclair & Lindgren 1955). Two vapor heat methods were compared: the earlier method which required heating the fruit to 43.3°C over a period of 8 h and holding the fruit at this temperature for an additional 8 h and 45 min and the quick run-up which comprised treating the fruit with saturated air at 49°C until the center of the fruits reached 48.5°C, which was approximately 4 h for oranges (navel and Valencia), grapefruits, lemons, and avocados. The quick run-up did not injure avocados or citrus significantly more than the long vapor heat treatment, with the exception of grapefruits which showed 7.2 and 17.5% injury when exposed to the 16.75 h of vapor heat treatment at 43.3°C and the quick run-up at 49°C until the center of the fruit reached 48.5°C, respectively. Concurrently, Claypool & Vines (1956) studied the reaction of 45 cultivars of 12 species of California deciduous fruits to the two vapor heat treatments. Although precise data were not presented, vapor heat damaged all fruits, with greater damage being caused by the quick run-up treatment.

By the early 1950s, ethylene dibromide (EDB) fumigation was recognized as an efficient, easy to use, and inexpensive commodity treatment. Use of vapor heat declined rapidly.

In 1978 work on vapor heat treatment of green bell pepper infested with oriental fruit fly was initiated in Japan because fumigation with EDB and methyl bromide caused damage to this commodity (Sugimoto et al. 1983). The recommended treatment was to heat peppers at 43.9 ± 0.3°C and >90% RH at a density of 90 kg per cubic meter until the fruit center reached 43°C and maintain that temperature for 3 h. The treatment did not cause excessive damage to the commodity.

Vapor heat treatment of eggplants also was studied in Japan because of phytotoxicity problems with fumigants (Furusawa et al. 1984). Eggplant density in the treatment chamber was directly related to the time required for the center of the eggplants to reach 43°C (when treated at 43.9 ± 0.3°C) and inversely related to percentage mortality of melon flies infesting the fruits. Eggplants tolerated vapor heat.

Mangoes grown in Okinawa tolerated the vapor heat treatment required to kill melon fly. For shipment to the rest of Japan, mangoes at a density of 80 kg per cubic meter of chamber were subjected to vapor heat at 44 ± 0.3°C and >90% RH until the pulp center reached 43°C and remained there for 3 h (Sunagawa et al. 1987). Vapor heat treatment was approved for mangoes shipped from the Philippines to Japan in 1986 (Magda 1987, Merino et al. 1985). Esguerra et al. (1990) found that the vapor heat treatment for mangoes, elevating the temperature of the seed surface to 46°C for 10 min, caused internal breakdown of the inner mesocarp in 15 - 80% of the mangoes. All eggs and first instars of oriental fruit fly in mangoes in Thailand were killed using vapor heat treatments of 44°C for 3 h, 46°C for 1 h, or 48°C for 30 min (Maglente 1986). The treatment accelerated ripening and reduced disease incidence in the mangoes. Kuo et al. (1987) developed a vapor heat treatment against melon fly and oriental fruit fly infesting mangoes in Taiwan which consisted of exposing the fruits to 47.5°C and 100% RH until the pulp next to the seed reached ≥46.5°C for 30 min. Eggs and third instars of *Bactrocera tryoni* (Froggatt) in mangoes were slightly more difficult to kill with vapor heat at 47°C than were first and second instars (Heard et al. 1992).

Sunagawa et al. (1988) did not succeed in developing a vapor heat treatment for bitter momordica fruit infested with melon fly without damaging the fruit.

Vapor heat treatment was investigated for grapefruit infested with Caribbean fruit fly (Miller et al. 1989, 1991, Hallman et al. 1990). The recommended treatment time of 4 h and 30 min was much shorter than the 14 h of treatment time used 40 years earlier for grapefruit infested with Mexican fruit fly in Texas and Mexico. This difference is probably due to the greater efficiency of modern vapor heat machinery and the use of statistical analysis to more accurately estimate time needed to kill fruit flies. Earlier estimates probably involved considerable overkill.

Vapor heat treatments were developed for carambolas in Taiwan and the U. S. The treatment developed in Taiwan used 45°C air at 100% RH until the center of the carambolas reached ≥44.2°C for 2 h and 30 min (Kuo et al. 1989). In the U. S., Hallman (1990, 1991) recommended that carambolas be treated at 46 - 46.3°C until center temperatures reached a minimum of 46°C for 35 min.

Surface insects were eliminated from tropical cut flowers and foliage with a vapor heat treatment at 46.6 ± 0.1°C for 1 h without damaging many of the plant materials studied (Hansen et al. 1992).

Corcoran et al. (1993) found eggs of *Bactrocera cucumis* (French) considerably more resistant to vapor heat at 45°C than all three instars inside zucchini, *Cucurbita pepo* L. Exposure of infested zucchinis to a vapor heat treatment that consisted of a 2 h heat-up from 30 to 46.5°C followed by 30 min at 46.5°C gave 100% mortality of an estimated 178,219 *B. cucumis* eggs.

Vapor heat treatments currently listed in the USDA-APHIS-PPQ Treatment Manual are for citrus and mangoes infested with Mexican fruit fly (14 h at 43.3°C) and for bell pepper, eggplant, papaya, pineapple, tomato, and zucchini squash infested with Mediterranean fruit fly, oriental fruit fly, or melon fly (44.4°C until center of fruit reaches 44.4°C, then held at that temperature for 8 h and 45 min) (Anonymous 1985).

Forced Hot Air

A heated air treatment was developed against Mediterranean fruit fly, melon fly, and oriental fruit fly infesting Hawaii-grown papayas that used 40 - 60% RH during the entire treatment without experiencing deleterious effects to fruit quality (Armstrong et al. 1989, Hansen et al. 1990). The treatment heated the centers of papayas to 47.2°C using a four-stage treatment that forced 43, 45, 46.5, and 49°C air over the fruit. The first three sequential stages each required approximately 2 h to heat the fruit center to 41, 44, and 46.5°C. The last stage required less than 1 h to heat the papaya center to 47.2°C (Anonymous 1990). The forced hot air treatment was not detrimental to the quality of papayas. Commercial facilities for applying the forced hot air treatment to papayas have been designed (Williamson & Winkelman 1989).

Although "vapor heat" treatments with <100% RH have been researched in the past, the forced hot air treatment was considered different from the vapor heat treatment because the latter specifies RH near 100% (Anonymous 1985). Forced hot air entered the USDA-APHIS-PPQ manual as the high temperature forced air treatment and is approved for papayas from Hawaii (Anonymous 1990).

Sharp & Hallman (1992) killed an estimated 214,801 Caribbean fruit fly larvae with no survivors in carambolas treated with hot air at 47 ± 0.2°C until the fruit center reached 45.5°C. Forced hot air treated carambolas deteriorated more rapidly than untreated carambolas (Miller et al. 1990). Sharp (1993) developed a hot-air quarantine treatment for grapefruits infested with Caribbean fruit fly consisting of exposure at 48 ± 0.3°C until the center of the grapefruits reached ≥44°C. Early- and mid-season grapefruits were predicted to tolerate hot air treatments up to 49°C for 2 h or 48°C for 3 h; however, it may not be possible to treat late-season grapefruit at temperatures >47.5°C without adversely affecting juice quality (McGuire & Reeder 1992).

Mangan & Ingle (1992) recommended that size 8 - 14 mangoes be treated at 50°C air temperature until the seed-surface temperature at maximum pulp thickness reached 48°C to disinfest mangoes of West Indian fruit fly. Hot air at 48°C until the seed surface reached 46.1°C was found to provide quarantine security for mangoes infested with Caribbean fruit fly (Sharp 1992).

Forced hot air at 47°C, 55 - 60% RH for 15 min killed 100% of adult greenhouse thrips, *Heliothrips haemorrhoidalis* (Bouché), adult longtailed mealybug, *Pseudococcus longispinus* Targioni-Tozetti, and fifth instar lightbrown apple moth, *Epiphyas postvittana* (Walker), on the surface of Japanese persimmon, *Disopyros kaki* L., fruits without lowering fruit quality (Cowley et al. 1992). However, mortality of adult female twospotted spider mite, *Tetranychus urticae* Koch, with this treatment was <15%. Diapausing twospotted spider mites were harder to kill with forced hot air than non-diapausing mites (Waddell & Birtles 1992). At 46.7°C, 100% mortality of diapausing twospotted spider mite on nectarines occurred in 15 h and 30 min.

Equipment for Applying Heated Air Treatments

Modern equipment for applying vapor heat and forced hot air treatments has been described (Sugimoto & Sunagawa 1987, Williamson & Winkelman 1989, Gaffney & Armstrong 1990, Gaffney et al. 1990, Sharp et al. 1991) (Fig. 10.1). These machines are operated by computer and capable of varying moisture content of the air, temperature, air velocity, and air flow direction. The ideal humidity to maintain while treating fruit with forced hot air is one derived from a dew point slightly below the surface temperature of the fruit to keep moisture from condensing on the fruit. However, the dew point should be as close to the surface temperature of the fruit as possible to prevent excess water loss from the fruit. Water loss from the fruit is deleterious in two ways: (1) as water is evaporated from the fruit it carries heat with it, thus, slowing heating of the fruit, and (2) water loss may lower fruit quality. Figure 10.2 presents generalized equipment for heated air treatments.

One problem encountered with early vapor heat machinery was in forcing air through the fruit load. Air flow follows the path of least resistance. Virtually all early equipment for vapor heat treatment used a room with an entrance vent at the top and an exit vent in the floor. Although the stacked boxes of fruit were set over the exit vent, forcing exiting air through the boxes sitting on the vent, the entrance vent was not connected to the top of the fruit stack; thus, air was not forced evenly through the fruit load resulting in uneven heating of the fruit.

Another problem with early vapor heat units was inadequate measurement and control of humidity. Workers commonly assumed 100% RH during the treatments. Also, air flow rate was usually low in early vapor heat equipment. Low air flow rate was largely responsible for the long time spans required to heat fruits. Another effect of low air flow rate is the lower rate of fruit heating near the air exit vent compared to the air entrance. As saturated air passes through the fruit load, water vapor condenses on the fruit and the air also loses heat by convection resulting in less heat available to heat fruit farther along the airflow path. As unsaturated air passes over fruit it may evaporate moisture from the fruit, resulting in loss of heat from the fruit. Williamson & Winkelman

FIGURE 10.1. Facility for applying heated air treatments at the U.S. Department of Agriculture Subtropical Horticulture Research Station in Miami, Florida (Sharp et al. 1991). Large box on left holds fruit; computer controls variables and records data.

158

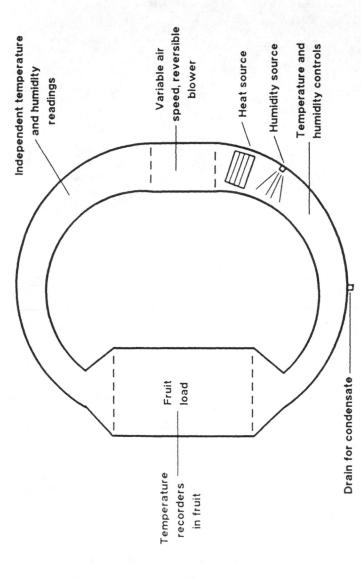

Independent temperature
and humidity
readings

Variable air
speed, reversible
blower

Heat source

Humidity source

Temperature and
humidity controls

Temperature
recorders
in fruit

Fruit
load

Drain for condensate

FIGURE 10.2 Schematized process for vapor heat and forced hot air treatments. Air cycles through fruit load and is replenished with heat and humidity as indicated by temperature and humidity controls. Ideally, air could alternately travel in either direction to prevent stratified heating of fruit. Treatments are usually ended when fruit center reaches pre-set temperature; therefore, temperature recorders should be placed in fruit most likely to reach pre-set temperature last.

(1989) solved the problem of reduction in heating along the heat path by reversing the air-flow direction halfway through the treatment.

Conclusions

Heated air quarantine treatments have been used for >60 years to prevent migration of mainly tephritid fruit flies. For the immediate future they will continue to be viable quarantine treatments. The difference between vapor heat and forced hot air is due to the APHIS definition of vapor heat having 100% RH. In actuality "vapor heat" treatments have been used which had <100% RH.

References

Akamine, E. K. 1966. Respiration of fruits of papaya (*Carica papaya* L., var. Solo) with reference to the effect of quarantine disinfestation treatments. J. American Soc. Hortic. Sci. 89: 231-236.

Anonymous. 1985. Animal and Plant Health Inspection Service. Plant protection and quarantine treatment manual, section III, part 9, pp. 1-3, Section VI-T106, p. 24. U.S. Government Printing Office. Washington, D.C.

_____. 1990. Animal and Plant Health Inspection Service. Plant protection and quarantine treatment manual, section VI-T102, pp 11-12. U.S. Government Printing Office. Washington, D.C.

Armstrong, J. W., J. D. Hansen, B.K.S. Hu & S. A. Brown. 1989. High-temperature, forced-air quarantine treatment for papayas infested with Tephritid fruit flies (Diptera: Tephritidae). J. Econ. Entomol. 82: 1667-1674.

Baker, A. C. 1952. "The Vapor-Heat Process," in *Insects: The Yearbook of Agriculture.* Pp. 401-404. Department of Agr. U.S. Government Printing Office. Washington, D. C.

Balock, J. W. & T. Kozuma. 1954. Sterilization of papaya by means of vapor-heat quick run-up. Special Report Ho-7. Entomol. Research Branch, Honolulu, Hawaii.

Balock, J. W. & D. F. Starr. 1945. Mortality of the Mexican fruitfly in mangoes treated by the vapor-heat process. J. Econ. Entomol. 38: 646-651.

Benlloch, M. 1934. Experiencias de desinfestación de frutas por el calor. Boletín de Patología Vegetal y Entomología Agrícola 7: 81-90.

Claypool, L. L. & H. M. Vines. 1956. Commodity tolerance studies of deciduous fruits to moist heat and fumigants. Hilgardia 24: 297-355.

Corcoran, R. J., N. W. Heather & T. A. Heard. 1993. Vapor heat treatment for zucchini infested with *Bactrocera cucumis* (Diptera: Tephritidae). J. Econ. Entomol. 86: 66-69.

Cowley, J. M., K. D. Chadfield & R. T. Baker. 1992. Evaluation of dry heat as a postharvest disinfestation treatment for persimmons. New Zealand J. Crop and Hortic. Sci. 20: 209-215.

Crawford, D. L. 1927. Investigation of Mexican fruit fly (*Anastrepha ludens* Loew) in Mexico. Monthly Bul. California Department Agr. 16: 422-445.

_____. 1929. Investigation of Mexican fruit fly (*Anastrepha ludens* Loew) in Mexico. Monthly Bul. State Plant Board of Fla. 12: 239-271.

Esguerra, E. B., S. R. Brena, M. U. Reyes & M.C.C. Lizada. 1990. Physiological breakdown in vapor heat-treated 'Carabao' mango. Acta Horticulturae 269: 425-434.

Food and Agricultural Organization (FAO). 1983. International Plant Quarantine Treatment Manual. FAO plant production and protection paper 50. FAO, Rome.

Furusawa, K., T. Sugimoto & T. Gaja. 1984. The effectiveness of vapor heat treatment against the melon fly, *Dacus cucurbitae* Coquillett, in eggplant and fruit tolerance to the treatment. Research Bul. Plant Protection Service (Japan) 20: 17-24 (in Japanese).

Gaffney, J. J. & J. W. Armstrong. 1990. High-temperature forced-air research facility for heating fruits for insect quarantine treatments. J. Econ. Entomol. 83: 1959-1964.

Gaffney, J. J., G. J. Hallman & J. L. Sharp. 1990. Vapor heat research unit for insect quarantine treatments. J. Econ. Entomol. 83: 1965-1971.

Hallman, G. J. 1990. Vapor-heat treatment of carambolas infested with Caribbean fruit fly (Diptera: Tephritidae). J. Econ. Entomol. 83: 2340-2342.

_____. 1991. Quality of carambolas subjected to postharvest hot water immersion and vapor heat treatments. HortScience 26: 286-287.

Hallman, G. J., J. J. Gaffney & J. L. Sharp. 1990. Vapor heat treatment for grapefruit infested with Caribbean fruit fly (Diptera: Tephritidae). J. Econ. Entomol. 83: 1475-1478.

Hansen, J. D., J. W. Armstrong, B. K. S. Hu & S. A. Brown. 1990. Thermal death of oriental fruit fly (Diptera: Tephritidae) third instars in developing quarantine treatments for papayas. J. Econ. Entomol. 83: 160-167.

Hansen, J. D., A. H. Hara & V. L. Tenbrink. 1992. Vapor heat: a potential treatment to disinfest tropical cut flowers and foliage. HortScience 27: 139-143.

Hawkins, L. A. 1929. Heat sterilization of citrus fruit. Plant Quarantine and Control Administration Service and Regulation Announcements

101: 225-227.

_____. 1930. Some factors affecting the heat sterilization of citrus fruit. Plant Quarantine and Control Administration Service Regulation Announcements 102: 25-27.

_____. 1932. Sterilization of citrus fruits by heat. Texas Citriculture 9: 7-8, 21-22.

Heard, T. A., N. W. Heather & P. M. Peterson. 1992. Relative tolerance to vapor heat of eggs and larvae of *Bactrocera tryoni* (Diptera: Tephritidae) in mangoes. J. Econ. Entomol. 85: 461-463.

Jones, W. W. 1939. The influence of relative humidity on the respiration of papaya at high temperatures. Proceedings, American Soc. Hortic. Sci. 37: 119-124.

_____. 1940. Vapor-heat treatment for fruits and vegetables grown in Hawaii. Hawaii Agric. Experiment Station Circular No. 16.

Jones, W. W., J. J. Holzman & A. G. Galloway. 1939. The effect of high-temperature sterilization on the Solo papaya. Hawaii Agric. Experiment Station Circular No. 14.

Koidsumi, K. 1936a. Heat sterilization of Formosan fruits for fruit flies (I) preliminary determinations on the thermal death points of *Chaetodacus ferrugineus* var. *dorsalis* Hendel and *C. cucurbitae* Coquillett. J. Soc. Tropical Agr. (Japan) 8: 157-165 (in Japanese).

_____. 1936b. Heat sterilization of Formosan fruits for fruit flies (II) results on "tankan" (*Citrus tankan* Hayata). J. Soc. Tropical Agr. (Japan) 8: 166-175 (in Japanese).

Koidsumi, K. & K. Shibata. 1936. On the velocity of heating and cooling of some fruits (a material for the control of fruit flies). J. Soc. Tropical Agr. (Japan) 8: 82-94 (in Japanese).

Kuo, L. S., C. Y. Su, C. Y. Hseu, Y. F. Chao, H. Y. Chen, J. Y. Liao & W. C. Huang. 1987. Vapor heat treatment for elimination of *Dacus dorsalis* and *Dacus cucurbitae* infesting mango fruits. Taiwan Bureau of Commodity Inspection and Quarantine, Ministry of Economic Affairs. Taipei.

Kuo, L. S., C. Y. Su, C. Y. Hseu, Y. F. Chao, H. Y. Chen, J. Y. Liao & C. F. Chu. 1989. Vapor heat treatment for elimination of *Dacus dorsalis* infesting carambola fruits. Taiwan Bureau of Commodity Inspection and Quarantine, Ministry of Economic Affairs. Taipei.

Latta, R. 1932. The vapor-heat treatment as applied to the control of narcissus pests. J. Econ. Entomol. 25: 1020-1026.

Mackie, D. B. 1931. Heat treatments of California fruits from the standpoint of compatibility of the Florida process. Monthly Bul. Dept. Agr. California 20: 211-218.

Magda, R. R. 1987. Vapor heat treatment for mango fruitfly control. Intl. Food Marketing and Technology. Nov. 35-36.

Maglente, E. E. 1986. Post harvest vapour heat treatment of mango to control Oriental fruit fly (*Dacus dorsalis* Hendel). M. S. thesis, Asian Institute of Technology, Bangkok, Thailand.

Mangan, R. L. & S. J. Ingle. 1992. Forced hot-air quarantine treatment for mangoes infested with West Indian fruit fly (Diptera: Tephritidae). J. Econ. Entomol. 85: 1859-1864.

McGuire, R. G. & W. R. Reeder. 1992. Predicting market quality of grapefruit after hot-air quarantine treatment. J. American Soc. Hortic. Sci. 117: 90-95.

Merino, S. R., M. M. Eugenio, A. U. Ramos & S. T. Hernandez. 1985. Fruitfly disinfestation of mangoes (*Mangifera indica* L. var 'Manila Super') by vapor heat treatment. Ministry of Agriculture and Food, Bureau of Plant Industry. Manila, Philippines.

Miller, W. R., R. E. McDonald, G. J. Hallman & M. Ismail. 1989. Phytotoxicity of hot water and vapor heat treatments to Florida grapefruit. Proceedings, International Conference on Technical Innovation in Freezing and Refrigeration of Fruits and Vegetables. Commissions C2, D1, D2, and D3. Pp. 207-212. University of California at Davis.

Miller, W. R., R. E. McDonald & J. L. Sharp. 1990. Condition of Florida carambolas after hot-air treatment and storage. Proceedings, Fla. State Hortic. Soc. 103: 238-241.

Miller, W. R., R. E. McDonald, G. Hallman & J. L. Sharp. 1991. Condition of Florida grapefruit after exposure to vapor heat quarantine treatment. HortScience 26: 42-44.

Seín, Jr., F. 1935. Heat sterilization of mangoes and guavas for fruit flies. J. Agr. University of Puerto Rico 19: 105-115.

Seo, S. T., B.K.S. Hu, M. Komura, C.Y.L. Lee & J. Harris. 1974. *Dacus dorsalis*: Vapor heat treatment in papayas. J. Econ. Entomol. 67: 240-242.

Sharp, J. L. 1992. Hot-air quarantine treatment for mango infested with Caribbean fruit fly (Diptera: Tephritidae). J. Econ. Entomol. 85: 2302-2304.

_____. 1993. Hot-air quarantine treatment for 'Marsh' white grapefruit infested with Caribbean fruit fly (Diptera: Tephritidae). J. Econ. Entomol. 86: 462-464.

Sharp, J. L. & G. J. Hallman. 1992. Hot-air quarantine treatment for carambolas infested with Caribbean fruit fly (Diptera: Tephritidae). J. Econ. Entomol. 85: 168-171.

Sharp, J. L., J. J. Gaffney, J. I. Moss & W. P. Gould. 1991. Hot-air treatment device for quarantine research. J. Econ. Entomol. 84: 520-527.

Sinclair, W. B. & D. L. Lindgren. 1955. Vapor heat sterilization of

California citrus and avocado fruits against fruit-fly insects. J. Econ. Entomol. 48: 133-138.

Sugimoto, T. & K. Sunagawa. 1987. Killing of fruit flies by vapor heat treatment of fruit. Plant Protection Service (Japan) 41: 124-128 (in Japanese).

Sugimoto, T., K. Furusawa & M. Mizobuchi. 1983. Effectiveness of vapor heat treatment against the oriental fruit fly, *Dacus dorsalis* Hendel, in green pepper and fruit tolerance to the treatment. Research Bul. Plant Protection Service (Japan) 19: 81-88 (in Japanese).

Sunagawa, K., K. Kume & R. Iwaizumi. 1987. The effectiveness of vapor heat treatment against the melon fly, *Dacus cucurbitae* Coquillett, in mango and fruit tolerance to the treatment. Research Bul. Plant Protection Service (Japan) 23: 13-20 (in Japanese).

Sunagawa, K., K. Kume, A. Ishikawa, T. Sugimoto & K. Tanabe. 1988. Efficacy of vapor heat treatment for bitter momordica fruit infested with melon fly, *Dacus cucurbitae* (Coquillett) (Diptera: Tephritidae). Research Bul. Plant Protection Service (Japan) 24: 1-5 (in Japanese).

Waddell, B. C. & D. B. Birtles. 1992. Disinfestation of nectarines of two-spotted mites (Acari: Tetranychidae). New Zealand J. Crop and Hortic. Sci. 20: 229-234.

Weddell, J. A. 1931. Experiments with the heat treatment of fly infested fruits. Queensland Agric. J. 36: 141-147.

Williamson, M. R. & P. M. Winkelman. 1989. Commercial scale heat treatment for disinfestation of papaya. American Society of Agric. Engineers Paper 89-6054.

11

Radio Frequency Heat Treatments

Guy J. Hallman and Jennifer L. Sharp

The radio frequency portion of the electromagnetic spectrum falls between the audio frequency and the infrared portions. It is the frequency range used for radio transmissions, from approximately 10 kHz to 100 GHz. Frequencies in that range have been used to heat objects by converting electromagnetic energy to heat energy. Microwaves (frequencies >1 GHz) are the most commonly used radio frequency waves for heating purposes. Radio frequency dielectric heating was developed before microwave heating and comprises frequencies <1 GHz. In any case, the heating process is identical at any wavelength in the radio frequency range.

Considerable research has been conducted using radio frequency heating to control pests of grains and nuts, and there are several reviews on this topic (Frings 1952, Thomas 1952, Nelson 1967, 1972, 1973, Kirkpatrick 1974). Hirose et al. (1970) have thoroughly studied the use of dielectric heating for controlling pests of cured tobacco. This chapter concentrates on the potential use of radio frequency heating to achieve quarantine security of pests of food plants. Relevant information on the use of radio frequency for nonquarantine pest control is included.

Headlee & Burdette (1929) pioneered the use of radio frequency to kill insects in 1928. Considerable work on the lethal effects of radio frequency electromagnetic fields to insects was performed in subsequent years (Davis 1933, Headlee & Jobbins 1938). Interpretations of the results of some of these earlier studies have been challenged as greater understanding developed of the factors affecting lethality due to radio frequency (Thomas 1952, Frings 1952). From this early research, important factors in achieving kill were identified, such as frequency, time of exposure, and field intensity (volts per meter).

Lethality of radio frequency is due to heating. Nonthermal lethal effects have not been conclusively demonstrated (Nelson 1973, Del Estal et al. 1986a). However, some questions remain as to additional modes of action (Nelson 1967).

The amount of heating that occurs when a material absorbs radio frequency energy can be estimated by:

$$dT/dt = 0.239P/cp$$

where dT is change in temperature (°C) in dt change in time (sec), P is watts per cubic centimeter, c is specific heat of the material, and p is density of the material (gram per cubic centimeter).

From the beginning of radio frequency research on insect kill, it was conjectured that discriminate heating of insects inside of agricultural commodities may be obtained by selected frequencies (Thomas 1952). Nelson & Charity (1972) recommended 10 - 100 MHz for selectively heating rice weevil, *Sitophilus oryzae* (L.), in hard red winter wheat. Dielectric heating at 40 MHz was more effective in killing pecan weevil, *Curculio caryae* (Horn), in pecan pieces than heating at 2,450 MHz (Nelson & Payne 1982). Nelson & Whitney (1960) found a differential heating factor of 1.2 - 2.3 for rice weevils in wheat. The factor decreased as the temperature increased. Frings (1952) did not feel that heating differentials could be practically exploited because they are modest, and other factors, such as orientation of pests inside the commodity and morphology of the pest and commodity, affect kill more significantly. In general, the radio frequency conductivity and the dielectric loss factor of the host must be lower than those of the pest for differential heating to be possible (Thomas 1952). Commercial microwave ovens operate at 2,450 MHz, a frequency at which little selective heating of insects is expected (Nelson 1987). However, speed of heating is directly correlated to frequency, and radio frequency heating will probably be most practical when used to heat the commodity as well as the pest to temperatures lethal to the pest without any benefit from differential heating. In this regard, radio frequency heating has an advantage over convective forms of heating commodities (hot water immersion and heated air) in that radio frequency heats commodities faster than convective heating. In any case, radio frequency heating would only be feasible in a small space because power requirements increase by the square of the distance from the radiating plate. For heating large lots of fruits they could be passed slowly on a conveyor.

Other variables affect mortality of insects by radio frequency heating.

The velocity of heating is inversely proportional to the mass and number of heated objects. Host of the insect may affect mortality. For example, granary weevils, *Sitophilus granarius* (L.), suffered higher mortality in corn than in wheat subjected to radio frequency heating (Nelson & Kantack 1966). Rice weevil adults survived radio frequency treatments better when mixed with small glass beads than with larger ones (Nelson et al. 1966). Frings (1952) found that appendages heated very rapidly and probably accounted for the faster kill of adults compared with larvae and pupae noted in some research (Webber et al. 1946, Del Estal et al. 1986b). At 25 MHz, large milkweed bugs, *Oncopeltus faciatus* (Dallas), 5 d old required twice the time to kill as bugs 30 d old. Mediterranean fruit fly, *Ceratitis capitata* (Wiedemann), adults 9 d old were slightly easier to kill than adults 2 d old, and pupae 1 d old were killed in half of the time required to kill pupae 4 or 8 d old at 9,000 MHz (Del Estal et al. 1986a,b). Male house flies, *Musca domestica* L., were considerably easier to kill than females (Frings 1952). The proportion of the vertical distance occupied by the objects being subjected to radio frequency and the distance between the two opposing plates of the radio frequency device also affect speed of heating. For example, with 15 and 30% of the vertical field occupied, and the voltage gradient maintained at 1,000 volts per centimeter, 34 and 21 sec, respectively, were required to kill adults of large milkweed bug (Frings 1952). At any given frequency, the voltage gradient between the plates greatly affected heating velocity. For example, adults of black blow fly, *Phormia regina* (Meigen), subjected to 25.4 MHz were killed in 30 and 8 sec with mean voltage gradients of 500 and 1,000 volts per centimeter, respectively (Frings 1952).

Frings (1952) believed that fruits could not be disinfested of fruit flies without damaging the fruits. Ample research with other methods of heating fruit has shown that many fruits can tolerate the heat necessary to kill fruit flies (Couey 1989). Nevertheless, the two examples of using radio frequency energy to kill insects inside of fruits have not been encouraging. Seo et al. (1970) used a microwave oven (2,450 MHz) to kill mango weevils, *Cryptorhynchus mangiferae* (F.), in mangoes. The mangoes were damaged by the temperatures generated (63°C resulted in 87% weevil mortality). Damage was lessened, but not eliminated, by treating the mangoes with repeated 10 - 15 sec bursts instead of continuous treatment (up to 60 sec). Hayes et al. (1984) subjected papayas infested with oriental fruit fly, *Bactrocera dorsalis* (Hendel), to microwave (2,450 MHz) until center temperatures reached 38 - 45°C followed by immersion in 48.7°C water for 8 - 20 min. Papaya center temperatures after microwave heating for fixed time periods varied within a range of 20°C. Therefore, heating papayas with microwave would require that some papayas be heated to 20°C above the target temperature to ensure that all papayas have been treated to the minimum target temperature, causing some papayas to be damaged.

Recommendations for Future Research

After initial optimism for use of radio frequency heat for killing pests in commodities, presently the treatment is not used commercially. High energy cost compared with other disinfestation methods, uneven heating, and damage to commodities have been the primary reasons why radio frequency has not been used commercially in insect control. Cost is probably not an objection to its use as a quarantine treatment; other heat treatments and cold storage are also energy intensive. Damage has been a concern especially in fresh commodities. In grains, pulses, nuts, dried fruits, cured tobacco, and similar commodities, radio frequency heat may have better potential as a quarantine treatment because these commodities are generally more resistant to heat spoilage than fresh commodities. However, damage to seeds intended for planting may occur (Nelson & Payne 1982, Crocker et al. 1987). Although the possibilities of using radio frequency treatments for quarantine pests inside fresh commodities do not look promising, research should be conducted using different frequencies, intermittent exposures and other methods to achieve quarantine security without damaging the commodities.

The combination of radio frequency with other quarantine treatments should be investigated. The combination might synergistically cause higher insect mortality than the additive effects of individual treatments. Tilton et al. (1972) achieved higher than expected mortality (85 - 91%) of Angoumois grain moth, *Sitotroga cerealella* (Olivier), using gamma irradiation (100 Gy) plus radio frequency (2,450 MHz for 25 sec).

References

Couey, H. M. 1989. Heat treatment for control of postharvest diseases and insect pests of fruits. HortScience 24: 198-202.

Crocker, R. L., D. L. Morgan & M. T. Longnecker. 1987. Effects of microwave treatment of live oak acorns on germination and on *Curculio* sp. (Coleoptera: Curculionidae) larvae. J. Econ. Entomol. 80: 916-920.

Davis, J. H. 1933. Radio waves kill insect pests. Scientific American 148: 272-273.

Del Estal, P., E. Viñuela, C. Camacho & E. Page. 1986a. "Biological Effects of Microwave Treatments on Pupae and Adults of *Ceratitis capitata* Wied.," in A. P. Economopoulos, ed., *Proceedings, II International Symposium on Fruit Flies/Crete*. Pp. 115-124. Athens: Elsevier.

Del Estal, P., E. Viñuela, E. Page & C. Camacho. 1986b. Lethal effects of microwaves on *Ceratitis capitata* Wied. (Dipt., Trypetidae) influence of development stage and age. J. Applied Entomol. 102: 245-253.

Frings, H. 1952. Factors determining the effects of radio-frequency electromagnetic fields on insects and materials they infest. J. Econ. Entomol. 45: 396-408.

Hayes, C. F., H.T.G. Chingon, F. A. Nitta & W. J. Wang. 1984. Temperature control as an alternative to ethylene dibromide fumigation for the control of fruit flies (Diptera: Tephritidae) in papaya. J. Econ. Entomol. 77: 683-686.

Headlee, Y. J. & R. C. Burdette. 1929. Some facts relative to the effect of high frequency radio waves on insect activity. J. New York Entomological Soc. 37: 59-64.

Headlee, Y. J. & D. M. Jobbins. 1938. Progress to date on studies of radio waves and related forms of energy for insect control. J. Econ. Entomol. 31: 559-563.

Hirose, T., I. Abe, R. Sugie, F. Miyanowaki & M. Kono. 1970. Studies on dielectric heating of tobacco: IV, on killing of tobacco injurious insects and an effect on filling capacity of tobacco shreds by microwave heating. Japan Monopoly Corporation Central Research Institute Scientific Papers 112: 75-87 (in Japanese).

Kirkpatrick, R. L. 1974. "The Use of Infrared and Microwave Radiation for the Control of Stored-product Insects," in *Proceedings, 1st International Working Conference on Stored Product Entomology*. Pp. 431-437. Savannah, Georgia.

Nelson, S. O. 1967. "Electromagnetic Energy," in W. W. Kilgore & R. L. Doutt, eds., *Pest Control--Biological, Physical, and Selected Chemical Methods*. Pp. 89-149. New York: Academic Press.

_____. 1972. Possibilities for controlling stored-grain insects with RF energy. J. Microwave Power 7: 231-239.

_____. 1973. Insect-control studies with microwaves and other radiofrequency energy. Bul. Entomological Soc. America 19: 157-163.

_____. 1987. Potential agricultural applications for RF and microwave energy. Transactions, American Soc. Agric. Engineers 30: 818-822, 831.

Nelson, S. O. & L. F. Charity. 1972. Frequency dependence of energy absorption by insects and grain in electric fields. Transactions, American Soc. Agric. Engineers 15: 1099-1102.

Nelson, S. O. & B. H. Kantack. 1966. Stored-grain insect control studies with radio-frequency energy. J. Econ. Entomol. 59: 588-594.

Nelson, S. O. & J. A. Payne. 1982. Pecan weevil control by dielectric heating. J. Microwave Power 17: 51-55.

Nelson, S. O. & W. K. Whitney. 1960. Radio-frequency electric fields for stored-grain insect control. Transactions, American Soc. Agric. Engineers 3: 133-137, 144.

Nelson, S. O., L. E. Stetson & J. J. Rhine. 1966. Factors influencing effectiveness of radio-frequency electric fields for stored-grain insect control. Transactions, American Soc. Agric. Engineers 9: 809-815.

Seo, S. T., D. L. Chambers, M. Komura & C.Y.L. Lee. 1970. Mortality of mango weevils in mangoes treated by dielectric heating. J. Econ. Entomol. 63: 1977-1978.

Thomas, A. M. 1952. Pest control by high-frequency electric fields--critical résumé. Technical Report W/T 23, British Electrical and Allied Industries Research Association, Surrey, England. 40 pp.

Tilton, E. W., J. H. Brower, G. A. Brown & R. L. Kirkpatrick. 1972. Combination of gamma and microwave radiation for control of the Angoumois grain moth in wheat. J. Econ. Entomol. 65: 531-533.

Webber, H. H., R. P. Wagner & A. Pearson. 1946. High-frequency electric fields as lethal agents for insects. J. Econ. Entomol. 39: 487-498.

12

Controlled Atmospheres

Alan Carpenter and Murray Potter

Controlled atmospheres (CA) have been developed to control a range of insect taxa in stored grains (Banks & Annis 1990). The use of lowered oxygen and elevated carbon dioxide levels to reduce insect damage to grains and dried fruits such as dates can be traced to the beginnings of agriculture (De Lima 1990). Recently, atmospheric modification has been applied to insect control on horticultural produce (Klag 1986). A CA may comprise inert gases (helium, nitrogen), low oxygen levels (2% or less), high carbon dioxide (5 - 60%), or combinations (Bailey 1957, Tunc 1983, Tunc & Navarro 1983). CA can be phytotoxic to horticultural produce, although severity will be a function of time and temperature as well as atmospheric composition (Kader 1989, Ke & Kader 1992). The literature on CA disinfestation reflects pest problems. In New Zealand there are many quarantine pests, and exporters target northern hemisphere gourmet markets and do not want to use methyl bromide to meet quarantine requirements. A great deal is known about CA control of insects such as New Zealand wheat bug, *Nysius huttoni* White, New Zealand flower thrips, *Thrips obscuratus* (Crawford), endemic leafrollers, *Ctenopseustis* species and *Planortortrix* species, the introduced light brown apple moth, *Epiphyas postvittana* (Walker), mealybugs, *Pseudococcus* species, codling moth, *Cydia pomonella* (L.) and green peach aphid, *Myzus persicae* (Sulzer) (Waddell et al. 1988, 1990, Whiting et al. 1991, 1992a, b, Batchelor 1992). These pests concern the Ministry of Agriculture, Japan, for kiwifruit, asparagus and stone fruits, and the United States Department of Agriculture authorities for apples, kiwifruit, asparagus, stone fruit, strawberry, and blueberry. CA research has focused on protecting crops from storage pests. This is especially true of dried fruits and nuts. Information derived for storage protection is also valuable for developing quarantine applications. This review combines information from a diverse range of sources and integrates it to provide a base for further applied work on the use of CA as a quarantine treatment.

Mode of Action

If an insecticidal treatment is to be applied across many pest taxa and use patterns, its mode of action must be understood. Friedlander (1983) suggested that CA has several, unrelated, effects on insect biochemistry:

1. More than 10% carbon dioxide stops the anaplerotic production of nicotinamide adenine dinucleotide (reduced form) [NADPH] which reduces the mixed function oxidases involved in much insecticide detoxification.
2. The energy charge is reduced, slowing the velocity of processes that require or produce adenosine triphosphate (ATP).
3. Production of glutathione, used in the detoxification of methyl bromide and related fumigants, is reduced.
4. High levels of carbon dioxide inhibit the regeneration of choline to acetylcholine.

Friedlander (1983) concluded that carbon dioxide has multiple sites of action. There are numerous accounts of the synergistic effect of carbon dioxide on fumigants suggesting that detoxification systems are affected (Carlson 1967).

Fleurat-Lessard (1990) reviewed the effects of elevated carbon dioxide and reduced oxygen on insects. He suggested that toxic oxygen levels are between 0.9 - 5%, with variation related to species, life stage, temperature and relative humidity (RH). Survival may depend on the ability of the test species to accumulate glycolytic products, reduce metabolic rate and restrict water loss. A possibility also exists that reduced oxygen affects synaptic efficiency.

The observation that high oxygen levels (20 - 50%) with high (50%) carbon dioxide caused mortality faster in grain beetles than atmospheres with less than atmospheric oxygen levels (20%) deserves more attention (Fleurat-Lessard 1990, Fleurat-Lessard & Le Torc'h 1991).

Various levels of carbon dioxide, nitrogen, and oxygen affect insect weight loss and survival (Navarro 1978, Jay & Cuff 1981, Jay 1983). Presumably, most weight loss is from the loss of water vapor, which leads to rapid mortality.

The few studies of the interactions between CA components suggest more studies on mode of action are needed to determine the roles and importance of low oxygen and elevated carbon dioxide in insect mortality (Soderstrom et al. 1986, Potter et al. 1993). Some insect species are tolerant of high carbon dioxide, and generally pupae are tolerant suggesting mechanisms exist that could allow resistance to high carbon dioxide to develop (Paton & Creffield 1987). Resistance to CA has been induced in granary weevil (Bond & Buckland 1979), but populations of tobacco moth, *Ephestia elutella* (Hübner), and red flour beetle, *Tribolium castaneum* (Herbst), that were resistant to malathion were not resistant to CA (Navarro 1978).

It seems likely that the internal pH of the insect can be affected by high carbon dioxide levels. The periodicity of carbon dioxide expiration is controlled by the spiracle cycle, and may also be related to the ability of carbonic anhydrase to convert carbon dioxide to bicarbonate (Miller 1974, Wigglesworth 1983).

One aspect of mode of action of CA that has not been explored is the simple physical impact of carbon dioxide on insects. Carbon dioxide dynamics are controlled in insects to minimize the impact of carbon dioxide on tissues and to prevent excessive loss of water. It seems likely that as external carbon dioxide levels rise above those of waste gas from the spiracle it will become increasingly difficult for the insect to rid itself of carbon dioxide. A build up of carbon dioxide in insect tissue may affect a wide range of biochemical processes, as reported for plant cells (Kubo et al. 1989, Li & Kader 1989).

Fleurat-Lessard (1990) recognized the biotic environment (food, species, life stage, age, weight, sex), physiological status (water balance, axon membrane permeability, activity, respiration rate, spiracle management, triglyceride levels), and the physical environment (temperature, RH, carbon dioxide, oxygen, time), as the important parameters affecting the susceptibility of insects to CA. Perhaps the quantitative, if not the qualitative, mode of action of CA would be affected by the habitat of the insect. Xeric stored-product insects are adapted to high carbon dioxide, low oxygen, and low humidity, rather different from the high humidity and normal atmosphere encountered by fruit infesting insects (Fleurat-Lessard 1990). Mitz (1979) and Nicolas & Sillans (1989) have discussed the dynamics of carbon dioxide on physiological systems. They concluded that oxygen limits respiration only. No data adequately explain the interactions between oxygen and carbon dioxide in CA efficacy. Is there synergism or are the effects simply additive?

A number of workers have identified sublethal effects of CA treatments. Soderstrom et al. (1991) found treatment with up to 60% carbon dioxide reduced fecundity and caused some mortality of adult codling moth. Lum et al. (1973) found that adult Indianmeal moth, *Plodia interpunctella* (Hübner), could lay viable eggs after they were dead, leading to reinfestation. Treatment of Indianmeal moth adults with 96% carbon dioxide for 1 h extended their lives and reduced egg number and viability (Lum & Phillips 1972, Lum & Flaherty 1972). Understanding how sublethal effects of CA occur may help understand the processes leading to mortality.

Insect and Acari Pest Groups

Tephritidae

The tephritid fruit flies are the world's most important quarantine pests. The potential of CA for fruit fly control was indicated by studies of carbon dioxide anaesthesia of insects (Sherman 1953). Benschoter et al. (1981) found 100% nitrogen and carbon dioxide concentrations of 40, 60, 80 and 100% toxic to eggs and larvae of Caribbean fruit fly, *Anastrepha suspensa* (Loew). Lethal effects were not proportional to concentration, suggesting a threshold for the effects of carbon dioxide. They concluded that despite carbon dioxide being slower acting than conventional fumigants, it was sufficiently active for practical use. In a detailed study of elevated carbon dioxide and reduced oxygen on egg and larval mortalities of the Caribbean fruit fly, oxygen levels (2, 10, or 20%) did not affect survival (Benschoter 1987). Treatments were more effective at 15.6 than 10°C. Every 10% increase in carbon dioxide level gave a 2% increase in mortality. The level of carbon dioxide needed for 100% mortality decreased by 20% for each additional day of exposure.

Apple maggot, *Rhagoletis pomonella* (Walsh), mortality in 3% oxygen and 2 - 8% carbon dioxide at 0 - 3°C was no different from that in cold storage (0 - 3°C) in air (Glass et al. 1961). Kosittrakun & Richardson (1993) found complete mortality of eggs and larvae after 35 d storage at ambient temperatures in air, after 24 d of storage at 0°C in air, and after 7 d storage at 20°C in nitrogen. Ali Niazee et al. (1989) found storing apple maggot eggs and larvae in 100% nitrogen at 20°C caused 100% mortality after 7 - 8 d. 'Granny Smith' apples and 'Anjou' pears did not develop off-flavors after 24 d, but both 'Golden Delicious' apples and 'Bartlett' pears developed off-flavors after 12 - 21 d in nitrogen at 0°C. Insecticidal nitrogen (100%) or low oxygen atmospheres (0.3%) for 7 d had no permanent effect on the sensory or physiological qualities of 'Golden Delicious' apples, 'Blue Jay' blueberries, 'Amity' red raspberries, 'Marion' blackberries and 'Italian' plums (M. Kosittrakun & D. G. Richardson, personal communication).

Treatment of blueberry maggot, *Rhagoletis mendax* Curran, with oxygen at 2 or 5% and levels of carbon dioxide from 25 - 95%, plus a treatment of 100% carbon dioxide, all at 5 or 21°C, for 24 or 48 h, showed that the oxygen level had a consistent effect on insect mortality (Prange & Lidster 1992). As carbon dioxide level increased, maggot mortality increased. Treatment at 21°C was more effective than it was at 5°C. The 100% carbon dioxide treatment was less effective than treatments containing 45 - 95% carbon dioxide (Prange & Lidster 1992). In blueberry fruit mortality of the blueberry maggot after 48 h treatment with 53% carbon dioxide in air at 25°C was 91% and was associated with a slight beneficial retention of fruit firmness over storage (Prange & Lidster 1992).

Jessup (1990) artificially infested 'Delica' squash with Queensland fruit fly eggs, *Bactrocera tyroni* (Froggatt), and stored them in air, 5% oxygen and 5% carbon dioxide or in pure nitrogen, at 10°C for three weeks. None of the atmospheres affected fruit fly survival. Fruit stored in nitrogen developed off-flavors.

Despite the importance of tephritid fruit flies as quarantine pests, data are scarce on the effects of CA.

Tortricidae

Within Tortricidae there are many insects of quarantine importance. The most significant species is codling moth. Many leafrollers are also important pests.

Mortality of both diapausing and nondiapausing fifth instar codling moth was complete after 48 h exposure to 95% carbon dioxide (Gaunce et al. 1982). After 33 h the larvae were either dead or moribund. In standard CA storage for apples (3% each of carbon dioxide and oxygen) the least resistant codling moth life stages were eggs and 0 - 13 d old larvae which died within 30 d (Moffitt & Albano 1972). Diapausing larvae required 133 d for complete mortality in a 0.8 - 1.6% CA of carbon dioxide and 2.2 - 3% oxygen (Toba & Moffitt 1989, 1991). In a large-scale test (1.5 - 2% oxygen and <1% carbon dioxide) nondiapausing codling moth larvae all died within 13 weeks (91 d) (Toba & Moffitt 1989, 1991). Thus 91 d commercial CA storage of apples would be an effective quarantine treatment for nondiapausing codling moth (Toba & Moffitt 1989, 1991).

Soderstrom et al. (1990) tested 60% carbon dioxide in air (8% oxygen) and a combustion product atmosphere (10% carbon dioxide and 0.5% oxygen) at 60 and 95% RH on codling moth. The lethal time was longer at the higher RH for each CA, except for eggs which were not affected by RH. High carbon dioxide was faster acting than low oxygen on all stages except diapausing larvae. In order of decreasing susceptibility to carbon dioxide, codling moth stages were egg, adult, pupae, larvae, and diapausing larvae. For low oxygen, the adults and larvae were reversed in order. Fresh produce needs to be held at around 95% RH; therefore, when diapausing larvae are present, the LT_{95} will be 25 d in 60% carbon dioxide. For nondiapausing life stages storage for 19 d in low oxygen or 9 d in 60% carbon dioxide would achieve LT_{95}.

Soderstrom et al. (1991) carried out an intensive study of CA control of codling moth. They found that LT_{95} values for eggs fell from 3.55 - 1.39 d as carbon dioxide level rose from 20 to 100% in air. In pure nitrogen, eggs had a LT_{95} of 1.16 d and in 0.5 to 5% oxygen (in nitrogen) eggs had LT_{95} values of 1.73 - 24.4 d. Thus the fastest acting CA for codling moth eggs was pure nitrogen followed by 40 - 100% carbon dioxide in air and then 0.5 - 2% oxygen. Levels of 20% carbon dioxide in air and 5% oxygen in nitrogen were rather less effective with LT_{95} values of 3.55 and 24.4 d respectively. When this work was extended to studying interactive effects of oxygen and carbon dioxide CA on egg survival, a more complex data set resulted. When 0 and 20% carbon dioxide was tested with 0 or 0.5% oxygen at 0, 5, 15 and 25°C, it was found that at 0 and 5°C temperature high mortality occurred in air with LT_{95} values around 10 d. CA at these temperatures were more effective with LT_{95} values around 3 - 4 d. At 15°C the low oxygen treatment was the most effective (LT_{95} 5.45 d) and the addition of 40% carbon dioxide extended the survival of eggs (LT_{95} 6.89 d). At 25°C the carbon dioxide alone was most effective (LT_{95} 1.04 d); in combination with low oxygen there was some inhibition (LT_{95} 2.29 d), and low oxygen alone was intermediate (LT_{95} 2.04 d). A similar experiment with 20% carbon dioxide and 2% oxygen gave rather different results. At 0°C the atmospheres were effective; the carbon dioxide alone was the most effective treatment (LT_{95} 4.64 d), the carbon dioxide and oxygen was intermediate (LT_{95} 5.24 d), and the low oxygen alone (LT_{95} 8.84 d) was little different from air (LT_{95} 10.57). At 15°C the 20% carbon dioxide was the most effective CA, but at 25°C the low oxygen was most effective. It seems that 20% carbon dioxide CA mixes were more effective than those containing 40%. The variation of mortality with temperature underlines our lack of understanding of the mode of action of CA in insect mortality. It was only with carbon dioxide levels of 40 and 60% that adult codling moth died (Soderstrom et al. 1991).

Batchelor et al. (1985) tested second and third instar light brown apple moth, green headed leafroller, *Planotortrix excessana* (Walker), and brown headed leafroller, *Ctenopseustis obliquana* (Walker), in commercial apple storage (about 3% carbon dioxide and 3% oxygen). Complete mortality of all three species occurred within 90 d.

In a CA of 3% carbon dioxide and 3% oxygen, first, third and fifth instars of light brown apple moths died within 32 d (Dentener et al. 1990). Later tests concentrated on third instars as they were more tolerant to CA than first instars. Fifth instars are likely to be eliminated due to fruit culling prior to packing. Calculated LT_{99} values were 17.8, 21 and 18.1 d for first, third and fifth instars, respectively. Six treatments were used, four at 0.5°C and two at 2°C. In the four at 0.5°C the LT_{99} values ranged from 20.9 to 47.6 d. In the tests at 2°C the lethal time was 58.1 - 56.9 d. These data indicate that at 0.5°C CA storage of apples will control light brown apple moth in <60 d. However, if temperatures rise slightly the time needed, especially for LT_{99} will increase. Waddell et al. (1990) carried out further experiments using the same conditions against the first, third and fifth instars of light brown apple moth, green headed leafroller, and brown headed leafroller in apples. For light brown apple moth the third instar was the most tolerant to CA. The fifth instars of green headed leafroller and brown headed leafroller were most tolerant. After 60 d storage all instars were dead. Whiting et al. (1991, 1992a, b) found that reducing oxygen levels (0.4 - 2%) and raising temperature to 20°C had more effect on mortality than raising carbon dioxide levels from 1 to 20%. Codling moth was the most resistant, taking over 200 h to reach LT_{99} (fifth instar), but it took only 100 h to achieve LT_{99} for green headed leafroller, brown headed leafroller and light brown apple moth with a CA of 0.4% oxygen and 5% carbon dioxide at 20°C, which was the most effective combination of those tested.

Standard CA storage will completely control codling moth and leafroller caterpillars in 60 - 90 d in apples. Many commodities, such as walnuts, are held in CA for periods at least this long, making this method of disinfestation practicable and economic. Stone fruits are less suited to this type of CA, unless CA compositions can be found that are more effective. High carbon dioxide levels over short periods may have potential, if the crop effects can be minimized. Soderstrom et al. (1990) found RH to be an important variable and suggested that different codling moth life stages were affected differently, although as both apples and stone fruit are stored under high RH, the information may not be able to be used in practice.

Marucci & Moulter (1971) showed field reduction of oxygen tension in cranberry bays to one - two parts per million, which was less than half the normal values, led to double the normal winter leafroller mortality. Their data indicate that the postharvest control of cranberry leafrollers with similar oxygen levels may be possible.

Diaspididae

Scale insects have been difficult to kill with CA. Morgan (1967) found that storage of San Jose scale, *Quadraspidiotus perniciosus* (Comstock), in CA stores (2 - 3% carbon dioxide and 3% oxygen) reduced the time to 100% mortality by 20 d when compared with standard cold storage. CA treatments for San Jose scale are complicated as mortality varies with picking date, cultivar, and seasonal conditions (Morgan & Angle 1967). When apples infested with San Jose scale were held at about 90% carbon dioxide at 2, 12, or 22°C, temperature affected the efficacy of the fumigation (Morgan & Gaunce 1975). Complete mortality was attained in 1 - 2 d at 22°C and 3 d at 12°C. At 2°C only 10 - 22% mortality occurred. In a second test at 22°C (96% carbon dioxide and <1% oxygen) it only took 1 d to kill all San Jose scale.

Apple responses to 97% carbon dioxide for 3 d were cultivar specific with 'Golden Delicious' developing skin damage and internal browning, while 'Spartan' and 'Red Delicious' were unmarked. All three cultivars developed off-flavors (Gaunce et al. 1982).

In Europe and Canada there were no differences in San Jose scale mortality in CA (3.5% or 1.5% carbon dioxide and 1.5 or 3% oxygen) compared with standard cold storage (Dickler 1976, Ferrari 1976, Angerelli et al. 1986). European fruit scale (oystershell scale), *Quadraspidiotus ostreaeformis* (Curtis), was not affected by CA storage after 236 d (Morgan 1967).

Tompkins et al. (1989) used carbon dioxide at 1,500 kPa for 10 min to control greedy scale, *Hemiberlesia rapax* (Comstock), on kiwifruit. The data show mortality of second and third instar greedy scale was higher with higher pressure and duration than it was in cool storage alone. Lower pressures and longer exposures may be necessary to avoid induction of storage rots (Archibald et al. 1990).

Curculionidae

Studies of CA treatments of sweetpotato weevil, *Cylas formicarius elegantulus* (Summers), demonstrated the interactions between CA composition, storage temperature, storage time, condition of the crop, life stage of the insect and insect microhabitat (Delate & Brecht 1989, Delate et al. 1990). A CA (8% oxygen and 40% carbon dioxide, at 30°C) killed 100% of adults in 8 d. Higher carbon dioxide levels reduced the exposure time needed for 100% kill. Immature stages of the weevil were more difficult to kill. One week in 4% oxygen and 40% carbon dioxide killed 98.3%, and one week in CAs of 4% oxygen and 60% carbon dioxide, 2% oxygen and 40% carbon dioxide, and 2% oxygen and 60%

carbon dioxide (75% RH), killed all immature stages present in infested roots. Delate et al. (1990) concluded that CA offers a practical solution to meeting quarantine requirements for sweet potatoes potentially infested with immature or adult stages of sweetpotato weevil. Delate & Brecht (1989) found that the lethal CA of 2% oxygen and 60% carbon dioxide was phytotoxic to the crop but that 2 or 4% oxygen and 40% carbon dioxide and 4% oxygen and 60% carbon dioxide were not phytotoxic, providing the tubers were cured at 30°C and 95% RH for 4 - 7 d to allow any wounds to heal prior to treatment.

Oxygen (1%) had no effect on pecan weevil, *Curculio caryae* (Horn), mortality, but mortality of weevils both inside and outside the shell was 74 - 90% in atmospheres of 30% carbon dioxide and 1% oxygen after two to six weeks (Wells & Payne 1980). When the CA mix was 30% carbon dioxide and 21% oxygen, mortality rose quickly from 62% after one month, to 92 and 95% after two and three months, to the desired 100% after four months. High carbon dioxide controlled weevils, reduced fungal infection, and preserved flavor. Storage of plum curculio larvae, *Conotrachelus nenuphar* (Herbst), at 2°C in 5% carbon dioxide and 2.5 - 5% oxygen did not affect mortality (Glass et al. 1961).

Acarina

Survival of overwintering apple rust mite, *Aculus schlechtendali* (Nalepa), and European red mite, *Panonychus ulmi* (Koch), was reduced greatly by CA storage (Lidster et al. 1981). A CA of 1% carbon dioxide and 1% oxygen was more effective than 5% carbon dioxide and 3% oxygen. Lidster et al. (1984) found that atmospheres with 1% oxygen killed all eggs of European red mite. In a CA treatment of 5% carbon dioxide and 3% oxygen, 53% of the eggs on apple were killed. Mortality increased with temperature (0 - 7.5°C) and length of time in storage. Lidster et al. (1984) concluded that CA could be used as an effective quarantine treatment for apple rust mite and European red mite eggs overwintering on apples.

Little mortality of European red mite eggs and McDaniel spider mites, *Tetranychus mcdanieli* McGregor, occurred after five months of storage (Gaunce et al. 1982). Treatment at 2 d in 100% nitrogen, 100% carbon dioxide, or 60% carbon dioxide and 40% nitrogen controlled European red mite eggs on apples, but 60% carbon dioxide and 40% air was ineffective. Some McDaniel spider mites survived 7 d in 100% carbon dioxide. Treatment of diapausing two-spotted mites, *Tetranychus urticae* Koch, with carbon dioxide at 1,500 kPa for 10 min controlled 80% of the mites on kiwifruit but induced fruit rots (Tompkins et al. 1989, Archibald et al. 1990).

Grain mite, *Acarus siro* L., was controlled after 3 d in 100% carbon dioxide (Hughes 1943). Navarro et al. (1985) exposed newly emerged grain mite adults to 2 - 21% oxygen and 10 - 40% carbon dioxide at 75% RH and at 15 and 26°C. In air, mites survived 11 d at 15°C and 7 d at 26°C. In 2% oxygen complete mortality occurred in 3 d at 15°C; at the higher temperature 10% oxygen was effective in 5 d. At 15°C, 30% carbon dioxide killed all grain mites in 4 d. At 26°C, 20% carbon dioxide killed all grain mites in 3 d. Viability of eggs of mold mite, *Tyrophagus putrescentiae* (Schrank), was reduced with carbon dioxide (Stepien 1979). Up to 6 d of treatment with carbon dioxide was needed to control mold mite eggs, but larvae needed only 10 h at 25°C, rising to 16 h for the protonymphs and 20 h for the deutonymph (Stepien 1979). Mites can be controlled in CA.

Hemiptera and Thysanoptera

Storage of apples at 0.5°C in 3% carbon dioxide and 3% oxygen caused complete mortality of New Zealand wheat bug (Waddell et al. 1990). Potter et al. (1990) used response surface analysis to represent data from experiments in which three levels of carbon dioxide (0, 9, 18% all with 2% oxygen) and four temperatures (0, 7, 13, 20°C) were held for two weeks to study the interactive effects of carbon dioxide levels and temperature on survival of New Zealand wheat bug and longtailed mealybug, *Pseudococcus longispinus* Targioni-Tozzetti. Longtailed mealybugs were cold sensitive, and CA increased mortality at higher temperatures. Wheat bugs were warm-temperature sensitive, with CA becoming an important mortality factor at lower temperatures.

Lill & van der Mespel (1986) studied the impact of CA (7% oxygen and 8% carbon dioxide) on New Zealand flower thrips and green peach aphid. Thrips mortality was 98% after 18 d compared with 30% in standard cold storage at 0°C. Mortality of alate and apterate green peach aphids was 73 and 89%, respectively. Complete kill of thrips and aphids was obtained after 12 - 21 d in 8% carbon dioxide and 7% oxygen storage or cold storage at 0°C (Carpenter & A.G. Stocker, unpublished data). Lowered RH and elevated carbon dioxide (60%) was found to give 100% mortality of New Zealand flower thrips and green peach aphids after 4.5 - 5.5 d without developing significant off-flavors in comparison with asparagus held under cold storage (Corrigan & Carpenter 1993). In 1990 this CA disinfestation method was tested by treating asparagus at 0 - 1°C in a CA of 60% carbon dioxide in air for 4.5 d, before air shipment to Japan, where no live insects were found. This trial shipment demonstrated that high carbon dioxide can be used as a commodity treatment (Carpenter & F. Bollen, unpublished data). Atmospheres of

24.4% carbon dioxide and 15.9% oxygen, 59.6% carbon dioxide and 8.4% oxygen, and 1 - 2% oxygen for 4 h at 21°C had no effect on green peach aphid infestations on head lettuce (Hartsell et al. 1979). Klaustermeyer et al. (1977) found that treatment of green peach aphids, corn earworms, *Helicoverpa zea* (Boddie), and cabbage looper, *Trichoplusia ni* (Hübner), on lettuce with atmospheres of 10, 30, 50, or 70% carbon dioxide with 5 or 21% oxygen at 2°C for 2 h, caused no insect mortality. Storage of lettuce infested with green peach aphids and cabbage loopers for 7 d under 5% oxygen, with or without 1% carbon dioxide, caused variable mortality, but always <80%.

Treatment of western flower thrips, *Frankliniella occidentalis* (Pergande), on strawberry with CA (50% carbon dioxide and 10% oxygen) did not cause any mortality in 4 h at 21°C (Aharoni et al. 1979). There was a narrow range (94 - 100%) in efficacy between treatments (about 2% oxygen with 0.2 - 2.5% carbon dioxide and 1.9 - 2.3% oxygen with 88.7 - 90.6% carbon dioxide), and the highest carbon dioxide treatment caused some off-flavors in strawberries (Aharoni et al. 1981).

Rose aphid, *Macrosiphum rosae* L., mortality after 24 h in 60% carbon dioxide was 97% (Stewart 1991). Third and fourth instars were the most resistant, apterate adults were intermediate, and first and second instars were least resistant.

Other Taxa

Pests of stored products are mostly cosmopolitan as a result of the antiquity of trade in these goods. Some have a wide host range including dried fruit and nuts, and some are carried on product into storage facilities from production areas. Pests from different primary environments may react differently to CA disinfestation (Fleurat-Lessard 1990, Soderstrom et al. 1990).

The stored products pests that are of quarantine importance are khapra beetle, *Trogoderma granarium* Everts, navel orangeworm, *Amyelois transitella* (Walker), and lesser grain borer, *Rhyzopertha dominica* (F.).

Khapra beetle was found to be relatively resistant to fumigation in walnuts with carbon dioxide and nitrogen, in comparison with merchant grain beetles, *Oryzaephilus mercator* (Fauvel), and almond moth, *Cadra cautella* (Walker) (Verma 1977, Verma & Wadhi, 1978). Khapra beetle larvae had LT_{99} values of 270 and 390 h, respectively, when treated with carbon dioxide and nitrogen. Pupae were less tolerant of carbon dioxide with a LT_{99} value of 105 h. Establishment of a CA in walnuts is complicated by sorption of carbon dioxide by walnut oils (Wells 1954).

Storey & Soderstrom (1977) found a CA of 21% oxygen and 9 - 9.5% carbon dioxide controlled navel orangeworm eggs 3 - 4 d old (LT_{95} at 8.4

h). Adults were more tolerant (LT$_{95}$ 34 - 39 h) and pupae were relatively resistant to the CA used (LT$_{95}$ 145 h). Bailey (1965) found that larvae of lesser grain borer adults died after 14 d in storage when oxygen fell below 4% and carbon dioxide levels rose above 45%.

There is a vast literature on CA control of stored products pests which provides a useful database for those researching CA as a quarantine tool. Recent reviews by Bailey & Banks (1980), De Lima (1990), Fleurat-Lessard (1990) and Soderstrom & Brandl (1990) give an overview of the work on stored product pests. Key CA factors are summarized for nonquarantine Coleoptera and Lepidoptera in Table 12.1.

There are significant differences between life stages in susceptibility to CA and there are major differences between taxa, both in overall susceptibility and ranking of life stage susceptibility. This is illustrated by data from a study of raisin and almond pests (Storey 1975). A generated atmosphere of <1% oxygen and 9 - 9.5% carbon dioxide in ambient conditions controlled larvae and adults of Indianmeal moth and almond moth, after 8 h. It took 24 h for eggs and pupae of Indianmeal moth and pupae of almond moth to be killed. Almond moth eggs died in 48 h. Angoumois grain moth, *Sitotroga cerealella* (Olivier), needed 24 h treatment to kill adults, 48 h treatment to kill eggs, 72 h to kill young larvae, 96 h to kill mid-aged larvae and 120 h to kill older larvae. An atmosphere of <0.5% oxygen and 12 - 14% carbon dioxide (at 27°C) caused complete mortality of Indianmeal moth, sawtoothed grain beetle, *Oryzaephilus surinamensis* (L.), adults and larvae, and red flour beetle adults in 48 h (Soderstrom & Brandl 1983, 1984). Red flour beetle and dried fruit beetle larvae (Nitidulidae) all died in 60 h, and raisin moths in 72 h. *Drosophila melanogaster* L. survived for 120 h. Decreasing the test temperature to 16°C greatly increased the time taken to achieve 100% mortality in all the species tested.

Shrink Wrapping

Shrink wrapping is used for the presentation and preservation of perishable crops. Shetty et al. (1989) found that no *D. melanogaster* L. survived in papaya and mangoes that were shrink wrapped for ≥72 h. The larvae of oriental fruit fly, *Bactrocera dorsalis* (Hendel), began to crawl to the surface of the papaya fruit within 30 min after wrapping. After 96 h in the wrap, mortality was 75 - 90%. Gould & Sharp (1990) found that shrink wrapping of mangoes controlled 99.95% of Caribbean fruit fly larvae after 15 d at 24 - 26°C giving an estimated probit 9 (99.9968%) mortality of 16.3 d. The mangoes began to decompose after 6 d. The authors concluded that shrink wrapping alone would not be a useful quarantine treatment but it may form a valuable adjunct to another treatment. Sharp (1990) found that the effect of wrapping grapefruit with Clysar films would provide probit 9 mortality of young Caribbean fruit fly larvae within 35 d. A few mature larvae survived, however, cut through the film, crawled out, and pupated.

Dentener et al. (1992) found that packaging of Japanese persimmons in films which caused a CA of about 0.5% oxygen and 5.3% carbon dioxide to develop within the film killed light brown apple moth larvae and mealybugs in 28 d. The advantage of this method is that it can be applied in transit.

Economics of the Application of CA

There have been few reports on the economics of the use of CA for disinfestation. A generated low oxygen atmosphere for raisin disinfestation was comparable in cost with the standard methyl bromide fumigation. The CA did not affect raisin quality (Guadagni et al. 1978, Soderstrom et al. 1984). Protection of grain against pests such as khapra beetle has been shown to be more effective with the use of CA in gas tight stores than with fumigants for peasant farmers in Senegal (Rouziere 1986a, b). Foscarini (1990) showed that a conventional railroad box car could maintain a CA atmosphere based on 60% or more carbon dioxide and control tobacco storage insects in transit (7 d).

CA has not been used for disinfestation of bulk dried fruits and tree nuts for two reasons (Soderstrom & Brandl 1990). Bulk stores are hard to seal, and the industry perceives CA disinfestation to be costly and slow in comparison to other approaches. Three factors that will change this situation are better sealing of stores, cheaper generation of CA, and increasing resistance of insects to phosphine. The facts that CA leaves no toxic residue in food and public demand for food that has not been treated with pesticides is increasing will result in pressure on the industry to use CA technology throughout the dried fruits and tree nuts industry, from bulk store to consumer pack (Soderstrom & Brandl 1990).

Discussion and Conclusions

CA disinfestation to meet quarantine requirements has been effectively demonstrated for of Tortricidae, Curculionidae, Miridae and Tephritidae that occur in or on fresh produce (Table 12.2). Extensive data exist for many taxa of insects that affect dried foods. Attempts to use CA treatments for Tephritidae and Acarina have had mixed success. Host product tolerance to treatment has often been a limiting factor. There are a range of pests and crops where CA has been shown to be an effective option for quarantine disinfestation but has not been taken up in practice.

Reasons for nonimplementation of CA disinfestation are not obvious (Armstrong 1985, 1992). Existing treatments may be cheaper or quicker, they may fit more readily with how an industry operates (size of packhouse, duration of produce storage, storage conditions, distance to market), or they may be more acceptable through being familiar. There is a lack of consistent data on a wide range of pests and on the effects of CA on produce. For CA to be used for disinfestation, more data are needed on a wider range of pests, particularly for the key groups Tephritidae, Diaspidae, and Acarina. Our trial export of CA disinfested asparagus to Japan in 1990 is apparently the only time CA has been used as a commercial quarantine treatment.

The historical development of CA was for quality maintenance and not for insect control (Kidd & West 1927). Recent reviews stress the likelihood of damage to apples from carbon dioxide in a CA and recommend 1 - 2% oxygen for an initial period, rising to 2 - 3% for the rest of the storage to retain aromatic qualities (Meheruik 1989, 1990). Carbon dioxide damage occurs in apples held in elevated carbon dioxide for extended periods; 8 - 12 months CA storage is now relatively common (Meheruik 1990). However, where elevated carbon dioxide has been used for insect suppression, storage periods in the CA are very much shorter (7 - 21 d). As the responses of pome fruit cultivars to CA vary (Meheruik 1989), it is likely that there may be varying cultivar responses to elevated carbon dioxide levels.

For produce storage, CA composition is held constant over the storage period (Kader 1989, Meheruik 1989, Saltveit 1989, 1990, Zagory & Reid 1989). Insect suppression with elevated carbon dioxide has been shown by this review to be an effective process, and there are indications that postharvest diseases may also be suppressed (Barkai-Golan 1990). Once the insect problem has been solved with elevated carbon dioxide, the CA in use in the storage could revert to the standard for the crop (e.g., for apples this would be about 3% each of carbon dioxide and oxygen) for the balance of the storage period. This approach may be called mixed-function CA to distinguish it from standard CA storage where the CA conditions are held at one level throughout the storage period. In mixed-function CA the composition could be varied over the storage period to achieve control of insects and disease, and to enhance final quality.

A major factor in determining the type of CA that might be used for quarantine purposes is the type of product being treated. We recognize four general classes of produce: (1) extremely perishable, can only be stored for a few days, (2) very perishable, can be stored for up to two months, (3) perishable, can be stored for up to one year, and (4) stable, can be stored longer than one year.

The perishability of a commodity will affect the CA procedures that can be carried out. Within a crop species, there may well be cultivar and seasonal differences in physiological tolerance and perishability that affect what types of CA can be used.

Bailey & Banks (1980) produced an excellent review of CA for stored products insect control. They identified key problems with the use of CA for insect control that are still relevant.

1. We do not know why some insects are sensitive to oxygen and why small changes in oxygen levels below 3 - 5% are crucial to insect mortality; we do not understand the effects of varying levels of carbon dioxide on insecticidal efficacy, and the cause of any synergism between carbon dioxide and oxygen is unknown.
2. The role of cold storage in the dynamics of CA for insect control varies from species to species and is not understood; no mechanism has been postulated for how low temperature provides some protection against the effects of low oxygen levels, or of how this protection can be overcome by raising carbon dioxide levels.
3. The causes of changes in relative susceptibility between species and between life stages of one species are unknown.
4. The ways in which CA affects water relations and spiracular rhythms of insects are poorly understood.
5. Little new data on mite suppression with CA has been published in the last 10 years, despite mites being of quarantine importance.
6. The physiological mechanisms leading to morbidity and latent effects of CA mixes on fecundity and growth are not understood.
7. No new research of induced resistance to CA exists.

Bailey & Banks (1980) suggested the extensive gaps in our knowledge were in part due to poor experimental designs. There are many data sets where trials were terminated before differences between treatments became clear, or simple designs prevent clear conclusions being drawn. Complex confounded factorial or response surface experimental designs are needed if clear conclusions are to be drawn (Potter et al. 1990, 1993, Prange & Lidster 1992). Lack of data on mite control with CA is a major problem if CA disinfestation is to be used when there are multi-species quarantine problems (Dentener et al. 1992). Thus despite the passage of time and the accumulation of data on CA disinfestation, the key areas of research identified in 1980 have still not been addressed (Bailey & Banks 1980).

The advantages of CA quarantine treatments are: (1) they can be used with little extra cost compared to traditional treatments, (2) equipment is minimal, and easily stored and readily transported, (3) there is little environmental impact, (4) treatments can be applied in transit, and (5) treatments can be devised that are compatible with quality maintenance of the commodities being treated.

If the quarantine needs of the international trade in fresh produce are to be met without the use of methyl bromide, researchers will need to be careful to ensure their work adds to our understanding of the use of CA to control insects. Little of the data published since the 1980 review has added to our understanding of CA as a quarantine treatment (Bailey & Banks 1980).

Data show that CA can be used for a quarantine treatment for many insects on many crops. Commercial scale trials are needed to show that CA disinfestation is effective, cost effective, meets consumers' needs for safe food, and can be accepted by regulatory authorities.

Acknowledgments

Helge van Epenhuijsen and Roberta Mayclair (Librarians, Levin Research Centre) provided invaluable help with literature retrieval. Our technician, Adrienne Stocker, helped with literature searching. Our colleagues Ian Stringer, Catherine Sinclair, Sandy Wright, and Carol Stewart provided stimulus and new ideas. We are indebted to David Swain, Ross Bicknell, Dave Leathwick, Ross Lill, and Sally Kerr (Levin Research Centre) who provided valuable comment on various drafts of this chapter. Jo Mulhane typed the manuscript.

Table 12.1. Summary of key factors in controlling non-quarantine stored product pests with CA treatments

Group	Key Factors	Reference
Coleoptera	Duration, Temperature, <5% O_2, 12% CO_2	Soderstrom & Brandl 1983, 1984
	Low O_2, elevated CO_2	Calderon & Navarro 1979, Tunc & Navarro 1983
	98-99% CO_2	Press & Harein 1967
	Duration, Temperature ≤4% O_2	Harein & Press 1988
	RH, ≥11% O_2	Pearman & Jay 1970
	Pure CO_2	Marzke & Pearman 1970
	Duration	Donahaye et al. 1988
	Temperature	White et al. 1988, 1990
	Temperature, Duration	Reichmuth 1986
	2% O_2 airtight	Oxley & Wickenden, 1963, De Lima 1990
	Reduced O_2	Bailey 1955, 1956, 1957, 1965
	RH, Temperature, Duration, >88% CO_2	Rameshbabu et al. 1991
Lepidoptera	Duration, Temperature <0.5% O_2, 12% CO_2	Soderstrom & Brandl 1983, 1984
	RH, O_2 >30% CO_2	Soderstrom et al. 1986
	≤6% O_2 >30% CO_2	Marzke et al. 1970
	Temperature, ≤5.7%, >44% CO_2	Tunc 1983, Tunc et al. 1982
	Temperature, Duration	Reichmuth 1986

Table 12.2 Summary of CA requirements for complete control of selected insect groups

Group	CA range tested	Temperature range Most resistant stage	Least resistant stage	Days to 100% mortality
Tephritidae	3% + oxygen 0-100% carbon dioxide 0-20°C		eggs and young larvae	10+
Tortricidae	0-5% oxygen 0-100% carbon dioxide 0-5°C	diapausing larvae	eggs and young larvae	2-91
Diaspididae	0-3% oxygen 0-90% carbon dioxide 0-5°C	none	none	1-200+
Curculionidae	2-4% oxygen 0-60% carbon dioxide 0-10°C	immature up to 120	adults	up to 120
Hemiptera	0-7% oxygen 0-18% carbon dioxide 0-20°C	none	adults	2 - 20+
Thysanoptera	0-7% oxygen 0-60% carbon dioxide 0-20°C	none	none	4 - 20
Acarina	2-21% oxygen 100% nitrogen 0-100% carbon dioxide 0-20°C	adults 11 - 150+	eggs	6+

References

Aharoni, Y., P. L. Hartsell, J. K. Stewart & D. K. Young. 1979. Control of western flower thrips on harvested strawberries with acetaldehyde in air, 50% carbon dioxide and 1% oxygen. J. Econ. Entomol. 72: 820-822.

Aharoni, Y., J. K. Stewart & D. G. Guadagni. 1981. Modified atmospheres to control western flower thrips on harvested strawberries. J. Econ. Entomol. 74: 338-340.

Ali Niazee, M. T., D. G. Richardson, M. Kosittrakum & A. B. Mohammad. 1989. "Non-insecticidal Quarantine Treatments for Apple Maggot Control in Harvested Fruit," in V.K. Fellman, ed., *Fifth International Controlled Atmosphere Research Conference*. Pp. 193-205. Wenatchee, Washington: University of Washington.

Angerilli, N.P.D., A. P. Gaunce & D. M. Logan. 1986. Some effects of post-harvest fumigation, controlled atmosphere storage, and cold storage on San Jose scale (Homoptera: Diaspididae) survival on two varieties of apples. Can. Entomologist 118: 493-497.

Archibald, R. D., A. R. Tompkins & M. E. Hopping. 1990. Killing mites and scale with carbon dioxide. New Zealand Kiwifruit J. September 1990: 16-17.

Armstrong, J. W. 1985. "Pest Organism Response to Potential Quarantine treatments," in *Proceedings, Regional Conference on Plant Quarantine Support for Agricultural Development*. Pp. 25-31. Association of South East Asian Nations, Plant Quarantine Centre and Training Institute. Serdang, Selangor, Malaysia.

_____. 1992. Fruit fly disinfestation strategies beyond methyl bromide. New Zealand J. Crop and Hortic. Sci. 20: 181-193.

Bailey, S. W. 1955. Airtight storage of grain; its effects on insect pests - I: *Calandra granaria* L. (Coleoptera: Curculionidae). Australian J. Agric. Research 6: 33-51.

_____. 1956. Airtight storage of grain; its effects on insect pests - II. *Calandra granaria* (small strain). Australian J. Agric. Research 8: 595-603.

_____. 1957. Air-tight storage of grain; its effect on insect pests - III. *Calandra granaria* (large strain). Australian J. Agric. Research 8: 595-603.

_____. 1965. Airtight storage of grain; ts effect on insect pests - IV. *Rhyzopertha dominica* (F) and some other Coleoptera that infest stored grains. Journal of Stored Products Research 1: 25-33.

Bailey, S. W. & H. J. Banks. 1980. "A Review of Recent Studies on the Effects of Controlled Atmospheres on Stored Product Pests," in J.

Shejbal, ed., *Controlled Atmosphere Storage of Grains.* Pp. 101-118. Amsterdam: Elsevier.

Banks, H. J. & P. C. Annis. 1990. "Comparative Advantages of High Carbon Dioxide and Low Oxygen Types of Controlled Atmospheres for Grain Storage," in M. Calderon and R. Barkai-Golan, eds., *Food Preservation by Controlled Atmospheres.* Pp. 94-122. Boca Raton, Florida: CRC Press.

Barkai-Golan, R. 1990. "Postharvest Disease Suppression by Atmospheric Modifications," in M. Calderon and R. Barkai-Golan, eds., *Food Preservation by Modified Atmospheres.* Pp. 238-264. Boca Raton, Florida: CRC Press.

Batchelor, T. A. 1992. Development of non-chemical disinfestation procedures in New Zealand using non-empirical, multi-disciplinary research. New Zealand J. Crop and Hortic. Sci. 20: 195-202.

Batchelor, T. A., R. L. O'Donnell & J. J. Roby. 1985. The efficacy of controlled atmosphere coolstorage in controlling leafroller species. Proceedings, New Zealand Weed and Pest Control Conference 38: 53-55.

Benschoter, C. A. 1987. Effects of modified atmospheres and refrigeration temperatures on survival of eggs and larvae of the Caribbean fruit fly (Diptera: Tephritidae) in laboratory diet. J. Econ. Entomol. 80: 1223-1225.

Benschoter, C. A., W. E. Knoop, & J. M. Owens. 1981. Toxicity of atmospheric gases to immature stages of *Anastrepha suspensa.* Fla. Entomologist 64: 543-544.

Bond, E. J. & C. T. Buckland. 1979. Development of resistance of carbon dioxide in the granary weevil. J. Econ. Entomol. 72: 770-771.

Calderon, M. & S. Navarro. 1979. Increased toxicity of low oxygen atmospheres supplemented with carbon dioxide on *Tribolium castaneum* adults. Entomologia Experimentalis et Applicata 25: 39-44.

Carlson, S. D. 1967. Temperature effect on mortality of confused flour beetles treated with carbon dioxide or nitrogen before fumigation. J. Econ. Entomol. 60: 1248-1250.

Corrigan, V. K. & A. Carpenter. 1993. The effects of treatment with elevated carbondioxide levels on the quality of asparagus. New Zealand J. Crop and Hortic. Sci. 21: (in press).

Delate, K. M. & J. K. Brecht. 1989. Quality of tropical sweet potatoes exposed to controlled-atmosphere treatments for postharvest insect control. J. of the American Soc. for Hortic. Sci. 114: 963-968.

Delate, K. M., J. K. Brecht & J. A. Coffelt. 1990. Controlled atmosphere treatments for control of sweet potato weevil. (Coleoptera: Curculionidae) in stored tropical sweet potatoes. J. Econ. Entomol. 83: 461-465.

De Lima, C.F.P. 1990. "Air Tight Storage: Principle and Practice," in M. Calderon and R. Barkai-Golan, eds., *Food Preservation by Modified Atmospheres*. Pp. 9-19. Boca Raton, Florida: CRC Press.

Dentener, P. R., B. C. Waddell & T. A. Batchelor. 1990. "Disinfestation of Light Brown Apple Moth: a Discussion of Three Disinfestation Methods," in B. B. Beattie, ed., *Managing Postharvest Horticulture in Australasia, Proceedings of The Australasian Conference in Postharvest Horticulture, 24-28 July 1989, Gosford, NSW, Australia.* Pp. 166-177. Australian Institute of Agric. Sci., Sydney, Australia. Occasional Publication No. 46.

Dentener, P. R., S. M. Peetz & D. B. Birtles. 1992. Non-chemical methods for the postharvest disinfestation of New Zealand persimmons. New Zealand J. of Crop and Hortic. Sci. 20: 203-208.

Dickler, E. 1976. Emfluss der kuhl - and CA lagerund von apfeln italienscher kerkunft anf mortalitat andfotilitat der San Jose schildlaus [(*Quadraspidiotus perniciosus* (Comstock)]. Redia 56: 401-416.

Donahaye, E., S. Navarro & M. Rindner. 1988. "The Influence of Different Treatments on the Disinfestation of Dates by Larvae of Nitidulid Beetles," *Progress Report for the Years 1985/87 of Stored Products Division*. Pp. 28-40. Agricultural Research Organisation: Bet Dagan, Israel.

Ferrari, R. 1976. Prove di sopravuivenza e fecondita del *Quadraspidiotus perniciousus* Comst. infestante mele conservate in frigorifero as atmosfera controllata. Redia 56: 391-400.

Fleurat-Lessard, F. 1990. "Effect of Modified Atmospheres on Insects and Mites Infesting Stored Products," in M. Calderon and R. Barkai-Golan, eds., *Food Preservation by Modified Atmospheres*. Pp. 21-38. Boca Raton, Florida: CRC Press.

Fleurat-Lessard, F. & J. M. Le Torc'h. 1991. Influence de la tenuer en oxygene sur la sensibilite de certains stadeo juveniles de *Sitophilus oryzae* et *S. granarius* au dioxyde de carbon. Entomologia Experimentalis et Applicata 58: 37-47.

Foscarini, U. 1990. Maggazzinaggio, transporto disinfestazione di tabacchi greggi in atmosphera controllata. Inform. Agrario 46 (Supl.13): 25-29.

Friedlander, A. 1983. "Biochemical Reflection on a Non-chemical Control Method. The Effect of Controlled Atmosphere on the Biochemical Processes in Stored Product Insects," in *Proceedings, Third International Working Conference on Stored Product Entomology*. Pp. 471-486. Manhattan, Kansas: Kansas State University.

Gaunce, A. P., C.V.G. Morgan & M. Meheruik. 1982. "Control of Tree Fruit Insects with Modified Atmospheres," in D. G. Richardson and M. Meheruik, eds., *Controlled Atmospheres for Storage and Transport of Perishable Agricultural Commodities*. Pp. 383-390. Oregon State

University School of Agriculture, Symposium Series 1. Beaverton, Oregon: Timber Press.

Glass, E. H., P. J. Chapman & R. M. Smoek. 1961. Fate of apple maggot and plum curculio larvae in apple fruits held in controlled atmosphere storage. J. Econ. Entomol. 54: 915-918.

Gould, W. P. & J. L. Sharp. 1990. Caribbean fruit fly (Diptera: Tephritidae) mortality induced by shrink-wrapping infested mangoes. J. Econ. Entomol. 83: 2324-2326.

Guadagni, D. G., E. L. Soderstrom & C. L. Storey. 1978. Effect of controlled atmosphere on flavour stability of almonds. J. Food Sci. 43: 1077-1080.

Harein, P. K. & A. F. Press. 1968. Mortality of stored-peanut insects exposed to mixtures of atmospheric gases at various temperatures. J. Stored Product Research 4: 77-82.

Hartsell, P. L., Y. Aharoni, J. K. Stewart & D. K. Young. 1979. Acetaldehyde toxicity to the green peach aphid on harvested head lettuce in high carbon dioxide or low oxygen atmospheres. J. Econ. Entomol. 72: 904-905.

Hughes, T. E. 1943. The respiration of *Tyroglyphus farinae*. J. Experimental Biology 20: 1-5.

Jay, E. 1983. "Imperfections in Our Current Knowledge of Insect Biology as Related to Their Response to Controlled Atmospheres," in B. E. Ripp, ed., *Controlled Atmosphere and Fumigation in Grain Storage*. Pp. 493-508. Amsterdam: Elsevier.

Jay, E. G. & W. Cuff. 1981. Weight loss and mortality of three stages of *Tribolium castaneum* (Herbst) when exposed to four controlled atmospheres. J. Stored Product Research 17: 117-124.

Jessup, A. J. 1990. "Postharvest Disinfestation Treatments; Recent Research Developments at NSW Agriculture and Fisheries Horticultural Postharvest Laboratory, Gosford," in B. B. Beattie, ed., *Managing Postharvest Horticulture in Australasia, Proceedings of The Australasian Conference in Postharvest Horticulture, 24-28 July 1989, Gosford, NSW, Australia..* Pp. 150-156. Australian Institute of Agric. Sci., Sydney, Australia. Occasional Publication No. 46.

Kader, A. A. 1989. "A Summary of CA Requirements and Recommendations for Fruits Other than Pome Fruits," in J. K. Fellman, ed., *International Controlled Atmosphere Research Conference, Fifth Proceedings*. Pp. 303-328. Wenatchee, Washington: University of Washington.

Ke, D. & A. A. Kader. 1992. Potential of controlled atmospheres for postharvest insect disinfestation of fruits and vegetables. Postharvest News and Information 3: 31N-37N.

Kidd, F. & C. West. 1927. A relation between the concentration of oxygen and carbon dioxide in the atmosphere, rate of respiration and length of storage life in apples. Great Britain Department of Scientific and Industrial Research Report of the Food Investigation Board 1925, 1926: 41.

Klag, N. G. 1986. "Use of CA for Quarantine Control of Insects on Fresh Fruits and Vegetables," in S. M. Blenkenship, ed., *Controlled Atmospheres for Storage and Transport of Perishable Agricultural Commodities, Horticultural Report No. 126*. Pp. 199-206. Department of Horticultural Science, North Carolina State University: Raleigh, North Carolina, U.S.A.

Klaustermeyer, J. A., A. A. Kader & L. L. Morris. 1977. "Effect of Controlled Atmospheres on Insect Control in Harvested Lettuce," in D. H. Dewey, ed., *Controlled Atmospheres for the Storage and Transport of Perishable Agricultural Commodities*. Pp. 203-204. Horticultural Report 28, Department of Horticulture, Michigan State University: East Lansing, Michigan.

Kosittrakun, M. & D. G. Richardson. 1993. Mortality of eggs and larvae of the apple maggot *Rhagoletis pomonella* (Walsh) (Diptera: Tephritidae) in 'Golden Delicious' apples as influenced by storage temperatures and nitrogen atmospheres. J.Econ. Entomol. 86: (in press).

Kubo, Y., A. Inaba & R. Nakamara. 1989. Effects of high carbon dioxide on respiration of various horticultural crops. J. of the Japanese Soc. of Hortic. Sci. 58: 731-736.

Li, C. & A. A. Kader. 1989. Residual effects of controlled atmospheres on postharvest physiology and quality of strawberries. J. of the American Soc. of Hortic. Sci. 114: 629-634.

Lidster, P. D., K. H. Sanford & K. B. McRae. 1981. Effects of modified atmosphere storage on overwintering populations of the apple rust mite and European red mite eggs. HortScience 16: 328-329.

_____. 1984. Effects of temperature and controlled atmosphere on the survival of overwintering populations of European red mite eggs on stored 'McIntosh' apples. HortScience 19: 257-258.

Lill, R. E. & G. J. van der Mespel. 1986. The effect of controlled atmosphere storage of asparagus on survival of insect passengers. Proceedings, New Zealand Weed and Pest Control Conference 39: 211-214.

Lum, P.T.M. & B. R. Flaherty. 1972. Effect of carbon dioxide on production and hatchability of eggs of *Plodia interpunctella* (Lepidoptera: Pyralidae). Annals, Entomological Soc. America 65: 976-977.

Lum, P.T.M. & R. H. Phillips. 1972. Combined effects of light and carbon dioxide on egg production of Indianmeal moth. J. Econ. Entomol. 65:

1316-1317.

Lum, P.T.M., B. R. Flaherty & R. H. Phillips. 1973. Fecundity and egg viability of stressed female Indianmeal moth, *Plodia interpunctella*. J. Georgia Entomological Soc. 8: 245-248.

Marucci, P. E. & H. J. Moulter. 1971. Oxygen deficiency kills cranberry insects. Cranberries March: 13-15.

Marzke, F. O. & G. C. Pearman. 1970. Mortality of red flour beetle adults and Indianmeal moth larvae in simulated peanut storages purged for short periods with carbon dioxide or nitrogen. J. Econ Entomol 63: 817-819.

Marzke, F. O., A. F. Press & G. C. Pearman. 1970. Mortality of the rice weevil, the Indianmeal moth, and *Trogoderma glabrum* exposed to mixtures of atmospheric gases at various temperatures. J. Econ. Entomol. 63: 570-574.

Meheruik, M. 1989. "CA Storage of Apples," in J. K. Fellman, ed., *International Controlled Atmosphere Research Conference, Fifth Proceedings*. Pp. 257-284. Wenatchee, Washington: University of Washington.

_____. 1990. Controlled atmosphere storage of apples: a survey. Postharvest News and Information 1: 119-121.

Miller, P. L. 1974. "Ventilation in Active and in Inactive Insects," in M. Rockstein, ed., *Principles of Insect Physiology (2nd ed)*. Pp. 367-390. New York: Academic Press.

Mitz, M. A. 1979. Carbon dioxide biodynamics: a new concept of cellular control. J. Theoretical Biology 80: 537-551.

Moffitt, H. R. & D. J. Albano. 1972. Effects of commercial fruit storage on stages of the codling moth. J. Econ. Entomol. 65: 770-773.

Morgan, C.V.G. 1967. Fate of San Jose scale and the European fruit scale (Homoptera: Diaspididae) on apples and prunes held in standard cold storage and controlled atmosphere storage. Can. Entomologist 99: 650-659.

Morgan, C.V.G. & B. J. Angle. 1967. Mortality of the San Jose scale (Homoptera: Diaspididae) on stored apples of different varieties and harvest dates. Can. Entomologist 99: 971-974.

Morgan, C.V.G. & A. P. Gaunce. 1975. Carbon dioxide as a fumigant against the San Jose scale (Homoptera: Diaspididae) on harvested apples. Can. Entomologist 107: 935-936.

Navarro, S. 1978. The effects of low oxygen tensions on three stored-product insect pests. Phytoparasitica 6: 51-58.

Navarro, S., O. Lider & U. Gerson. 1985. Response of adults of the grain mite *Acarus siro* L. to modified atmospheres. J. Agric. Entomol. 2: 61-68.

Nicolas, G. & D. Sillans. 1989. Immediate and latent effects of carbon dioxide on insects. Annual Review Entomol. 34: 97-116.

Oxley, T. A. & G. Wickenden. 1963. The effect of restricted air supply on some insects which infect grain. Annals of Applied Biology 51: 313-324.

Paton, R. & J. W. Creffield. 1987. The tolerance of some timber insect pests to atmospheres of carbon dioxide and carbon dioxide in air. International Pest Control 29: 10-12.

Pearman, G. C. & E. G. Jay. 1970. The effect of relative humidity on the toxicity of carbon dioxide to Castane *Tribolium castaneum* in peanuts. J. Georgia Entomological Soc. 5: 61-64.

Potter, M. A., A. Carpenter & A. Stocker. 1990. "Response Surfaces for Controlled Atmosphere and Temperature by Species: Implications for Disinfestation of Fresh Produce for Export," in B. B. Beattie, ed., *Managing Postharvest Horticulture in Australasia, Proceedings of The Australasian Conference in Postharvest Horticulture, 24-28 July 1989, Gosford, NSW, Australia.*. Pp. 183-190. Australian Institute of Agric. Sci., Sydney, Australia. Occasional Publication No. 46.

Potter, M. A., A. Carpenter, A. Stocker & S. Wright. 1993. Controlled atmospheres for the postharvest disinfestation of crops infested with adults of New Zealand Flower thrips (Thysanoptera: Thripidae). J. Econ. Entomol. 86: (in press).

Prange, R. K. & P. D. Lidster. 1992. Controlled atmosphere effects on blueberry maggot and lowbush blueberry fruit. HortScience 27: 1094-1096.

Press, A. F. & P. K. Harein. 1967. Mortality of *Tribolium castaneum* (Herbst) (Coleoptera: Tenebrionidae) in simulated peanut storages purged with carbon dioxide and nitrogen. J. Stored Product Research 3: 91-96.

Rameshbabu, M., D. S. Jayas & N.D.G. White. 1991. Mortality of *Cryptolestes ferrugineus* (Stephens) adults and eggs in elevated carbon dioxide and depleted oxygen atmospheres. J. Stored Product Research 27: 63-170.

Reichmuth, C. 1986. "Low Oxygen Content to Control Stored Product Insects," in E. Donahaye and S. Navarro, eds., *Proceedings, 4th Int. Work Conf. Stored-product Protection.* Pp. 194-207. Tel Aviv, Israel.

Rouziere, A. 1986a. Stockage des semences d'arachide decortiquees en atmospheres controllees. I Essais preliminaires 1979-1982. Oleagineux 41: 329-344.

_____. 1986b. Stockage des semences d'arachide descortiquees au atmosphere controllees. II Essais de prevulgarisation. Oleagineaux 41: 507-518.

Sharp, J. L. 1990. Mortality of Caribbean fruit fly immatures in shrink wrapped grapefruit. Fla. Entomologist 73: 660-664.

Sherman, M. 1953. Effects of carbon dioxide on fruit flies in Hawaii. J. Econ. Entomol. 46: 15-19.

Shetty, K. K., M. J. Klowden, E. B. Jang & W. J. Kochan. 1989. Individual shrink wrapping: a technique for fruitfly disinfestation in tropical fruits. HortScience 24: 317-319.

Soderstrom, E. L. & D. G. Brandl. 1983. "Modified Atmospheres for Postharvest Insect Control in Tree Nuts and Dried Fruits," in *Proceedings, Third International Working Conference on Stored Product Entomology*. Pp. 487-497. Manhattan, Kansas: Kansas State University.

_____. 1984. Low-oxygen atmosphere for postharvest insect control in bulk-stored raisins. J. Econ Entomol 77: 440-445.

_____. 1990. "Controlled Atmospheres for the Preservation of Tree Nuts and Dried Fruits," in M. Calderon and R. Barkai-Golan, eds., *Food Preservation by Modified Atmospheres*. Boca Raton, Florida: CRC Press.

Soderstrom, E. L., P. D. Gardner, J. L. Bartelle, K. N. de Lozano & D. G. Brandl. 1984. Economic cost evaluation of a generated low oxygen atmosphere as an alternative fumigant in the bulk storage of raisins. J. Econ. Entomol. 72: 457-461.

Soderstrom, E. L., B. E. Mackey & D. G. Brandl. 1986. Interactive effects of low-oxygen atmospheres, relative humidity, and temperature on mortality of two stored-product moths (Lepidoptera: Pyralidae). J. Econ. Entomol. 79: 1303-1306.

Soderstrom, E. L., D. G. Brandl & B. Mackey. 1990. Responses of codling moth (Lepidoptera: Tortricidae) life stages to high carbon dioxide or low oxygen atmospheres. J. Econ. Entomol. 83: 472-475.

_____. 1991. Responses of *Cydia pomonella* (L.) (Lepidoptera: Tortricidae) adults and eggs to oxygen deficit on carbon dioxide enriched atmospheres. J. Stored Product Research 27: 95-101.

Stepien, Z. 1979. "Effect of carbon dioxide on *Tyrophagus putrescentiae* (Schr.) (Acarina: Acaridae)," in E. Piffl, Ed., *Proceedings, Fourth International Congress of Acarology*. Pp. 249-255. Saalfelden am Steinernen (Austria) Budapest: Akademiai Kiado.

Stewart, C. A. 1991. The effects of controlled atmospheres on Macrosiphum rosae (L.), the rose aphid. Internal Report, Horticultural Research Centre, Levin, New Zealand.

Storey C. L. 1975. Mortality of three stored product moths in atmospheres produced by an exothermic inert atmosphere generator. J. Econ. Entomol. 68: 736-738.

Storey, C. L. & E. L. Soderstrom. 1977. Mortality of navel orangeworm in a low oxygen atmosphere. J. Econ. Entomol. 70: 95-97.

Toba, H. H. & H. R. Moffitt. 1989. "Controlled Atmosphere Storage as a Postharvest Quarantine Treatment for Codling Moth on Apples," in D.S. Reid (preparator), *International Conference on Technical Innovations in Freezing and Refrigeration of Fruits and Vegetables.* Pp. 213-214. Paris: International Institute Refrigeration.

_____. 1991. Controlled atmosphere cold storage as a quarantine treatment for non-diapausing codling moth (Lepidoptera: Tortricidae) larvae in apples. J. Econ. Entomol. 84: 1316-1319.

Tompkins, A. R., R. D. Archibald & M. E. Hopping. 1989. "Evaluation of Pressurised Carbon Dioxide and Refrigeration to Kill Scale and Mite Contaminants of Kiwifruit," in D.S. Reid (preparator), *International Conference on Technical Innovations in Freezing and Refrigeration of Fruits and Vegetables.* Pp. 220-229. Food Science and Technology, University of California, Davis.

Tunc, I. 1983. The effect of low oxygen- and high carbon dioxide-atmospheres on the eggs and larvae of *Plodia interpunctella.* Zeitschrift fur angewandt Entomologie 95: 53-57.

Tunc, I. & S. Navarro. 1983. Sensitivity of *Tribolium castaneum* eggs to modified atmospheres. Entomologia Experimentalis et Applicada 34: 221-324.

Tunc, I., C. Reichmuth & R. Wohlegemuth. 1982. A test technique to study the effects of controlled atmospheres on stored product pests. Z. ang. Entomol. 93: 493-496.

Verma, A. K. 1977. Effect of atmospheric gases on pest infestation during storage and on keeping quality of walnuts. Entomological Newsletter 7: 13-14.

Verma, A. K. & S. R. Wadhi. 1978. Susceptibility of walnut pests to carbon dioxide and nitrogen and effect of gas storage on keeping quality of walnut kernels. Indian J. Entomol. 40: 290-298.

Waddell, B. C., V. M. Cruikshank and P. R. Dentener. 1988. Postharvest disinfestation of *Nysius huttoni* on "Granny Smith" apples using controlled atmosphere storage. DSIR Entomology Division Internal Report, Auckland, New Zealand.

Waddell, B. C., P. R. Dentener & T. A. Batchelor. 1990. Time mortality responses of leafrollers exposed to commercial controlled atmosphere storage. Proceedings, New Zealand Weed and Pest Control Conference 43: 328-333.

Wells, A. W. 1954. Sorption of carbon dioxide by nut meats. Science 120: 188.

Wells, J. M. & J. A. Payne. 1980. Reduction of mycoflora and control of in-shell weevils in pecans stored under high carbon dioxide atmospheres. Plant Disease 64: 997-999.

White, N.D.G., D. S. Jayas & R. N. Sinha. 1988. Interaction of carbon dioxide and oxygen levels and temperature on adult survival and reproduction of *Cryptolestes ferrugineus* in stored wheat. Phytoprotection 69: 31-39.

_____. 1990. Carbon dioxide as a control agent for the rusty grain beetle (Coleoptera: Cucujidae) in stored wheat. J. Econ. Entomol. 83: 277-288.

Whiting, D. C., S. P. Foster & J. H. Maindonald. 1991. The effects of oxygen, carbon dioxide, and temperature on the mortality responses of *Epiphyas postvittana* (Lepidoptera: Tortricidae) J. Econ. Entomol. 84: 1544-1549.

_____. 1992a. Comparative mortality responses of four tortricid (Lepidoptera) species to a controlled atmosphere of 0.4% oxygen/5% carbon dioxide. J. Econ. Entomol. 85: 2305-2306.

Whiting, D. C., J. van den Heuvel & S. P. Foster. 1992b. The potential of low oxygen/moderate carbon dioxide atmospheres for postharvest disinfestation of New Zealand apples. New Zealand J. Crop and Hortic. Sci. 20: 217-222.

Wigglesworth, V. B. 1983. The physiology of insect tracheoles. Advances in Insect Physiology 17: 85-148.

Zagory D. & M. S. Reid. 1989. "Controlled Atmosphere Storage of Ornamentals," in J. K. Fellman, ed., *International Controlled Atmosphere Research Conference, Fifth Proceedings.* Pp. 353-358. Wenatchee, Washington: University of Washington.

13

Commodity Resistance to Infestation by Quarantine Pests

John W. Armstrong

International trade in agricultural commodities provides many avenues for quarantine pests to enter geographical locations where the pests do not occur. The prohibition of trade of commodities that are hosts for quarantine pests is acceptable if no quarantine treatment is available that ensures quarantine security. Often, quarantine treatments are used to disinfest commodities.

Many quarantine treatments are available and include heat and cold (Chapters 8, 9, 10, and 11), irradiation (Chapter 7), and fumigation (Chapter 5). Others showing promise are under study. Public concern about health hazards attributed to the use of pesticides, especially for commodities that are consumed, has led to increasingly tighter restrictions. Many pesticides have been banned, including ethylene dibromide (Ruckelshaus 1984). The chemical was used worldwide as a fumigation treatment to disinfest tephritid fruit flies from host fruits (Armstrong & Couey 1989). Today, the major emphasis and challenge in quarantine treatment research are to develop alternate treatments to using pesticides.

Nonhost or infestation-resistant commodities and cultivars, stages of maturity of hosts, and growing periods have received little attention as alternate approaches to quarantine, perhaps because of the difficulty in developing supporting data and the necessary reliance on inspection to maintain quarantine security. This chapter reviews and discusses examples whereby resistance or nonsusceptibility to infestation has allowed different commodities to enter marketing channels without quarantine treatment. The term, resistance, which usually implies a genetic attribute, is used herein to describe all factors that affect the susceptibility of commodities to infestation. Although most examples relate to tephritid fruit flies because of their economic and quarantine importance, the described concepts generally can be applied to all quarantine pests.

Host and Nonhost

A quarantine host is defined here as any commodity which can be naturally infested in the field at one or more of its growth stages. This may occur after harvest and processing in the case of stored-product quarantine pests, on or in a host where the pest can successfully complete its life cycle, and when the host is transported to an area where it does not already exist (Armstrong 1985). Stored-product quarantine pests, which usually infest commodities after harvest, generally complete their life cycles in the commodity. This definition of quarantine host overlooks commodities which are resistant to infestation during certain stages of growth or maturity.

Historically, host status was conferred upon commodities found with the respective quarantine pest on or in the commodity in the field, with the exception of stored products. The accurate determination of host status is often a difficult and controversial problem. Casual observations and poorly documented reports occasionally have led to quarantine requirements that have little or no objective validity (Couey 1983). For example, many published articles exist which list the hosts of Mediterranean fruit fly, *Ceratitis capitata* (Wiedemann), without any accompanying infestation data. Perhaps some recognized hosts of Mediterranean fruit fly for regulatory purposes are not biological hosts. Determination of the biological status of many reported hosts is difficult because of the scarcity of studies on host acceptance, recognition, selection, and life history information (Liquido et al. 1991).

Although locality records differ from host records, the importance of locality records to quarantine warrants brief mention. Although a commodity is a host of a quarantine pest, regulation is unnecessary if the pest is not present in the area where the commodity is grown. Erroneous locality records can lead to problems similar to those caused by erroneous host records. For example, Webster (1920) reported that mango weevil, *Cryptorhynchus mangiferae* (F.), a quarantine pest of mangoes, *Mangifera indica* L., was present in the Philippines. Because no quarantine treatment against mango weevil was available, Philippine mangoes could not be exported to the United States (U.S.) regardless that inspections never produced any mango weevil infestations. A large-scale study of mangoes sampled from the major mango producing areas in the Central Visayas showed that mango weevil was not present (Corey 1985). An interesting aspect of host status is the suitability to infestation by the quarantine pest, i.e., good hosts versus poor hosts. Tephritid fruit fly host lists often group hosts into different categories such as heavily or generally infested, occasionally infested, rarely infested, and infested only in the laboratory (USDA 1983). These categories were developed by a number of individuals who may have had divergent ideas regarding what constituted each host category. Anecdotal information rather than scientific data may have been used on occasion. Thus, these host categories are seldom recognized for the regulatory application of quarantine treatments. The primary goal of regulatory agencies is to keep quarantine pests out of their respective countries. Obviously, from a regulatory point of view, a host must be considered a host regardless of whether it is generally, occasionally, or only rarely infested. This philosophy requires the same efficacy data for treatments of good hosts and poor hosts, alike. For example, the United States Department of Agriculture, Animal and Plant Health Inspection Service, Plant Protection and Quarantine (USDA-APHIS-PPQ) requires efficacy data showing probit 9 (98.9968%) mortality at the 95% confidence limit, or no more than 3.2 survivors from a treated population of 100,000 (Baker 1939). Because of the requirement, use of the probit 9 concept for ensuring quarantine security has been questioned (Landolt et al. 1984, Baker et al. 1990, Harte et al. 1992, Chapters 3, 4), and is often difficult to use in the research and development of quarantine treatments. Development of efficacy data was relatively simple for methyl bromide treatment schedules for stone fruits infested by oriental fruit fly, *Bactrocera dorsalis* (Hendel), because nectarine, peach, and plum were easily infested and produced >1,200, >1,500 and >800 survivors, respectively, per kg of infested fruit (Armstrong et al. 1988). Conversely, the development of efficacy data may be difficult for fruits that are poor hosts, such as lychee, *Litchi chinensis* Sonnerat, which may produce only 0 - 50 survivors per kg of infested fruit.

Another aspect of host status is whether the host is a natural or an artificial host. A natural host is one that is infected (pathogens) or infested (pests) by natural means; an artificial host is one that is infected or infested by experimental means (Kahn 1989). Not all artificial hosts are known to be natural hosts. Not all natural hosts have been infected or infested artificially. Thus, a given host may be either a natural or an artificial host, or both. If a commodity can be only artificially infested with a quarantine pest, perhaps it should not be considered a host for quarantine purposes or require treatment (Couey 1983). For example, solanum fruit fly, *Bactrocera latifrons* (Hendel), was successfully reared on papaya in Hawaii following artificial infestation (Vargas & Nishida 1985a). However, papaya is not a host in nature for solanum fruit fly, which prefers solanaceous and curbitaceous fruits (Vargas & Nishida 1985b). It would be erroneous to require a quarantine treatment against solanum fruit fly in papaya exported to the U.S. mainland.

Another example is 'Smooth Cayenne' pineapple, *Ananas comosus* (L.). The pineapple cultivar can be artificially infested with melon fly, *Bactrocera cucurbitae* (Coquillett), and oriental fruit fly in the laboratory using severe infestation pressure and ideal holding conditions, or by oriental fruit fly when the fruit is overripe to the point of being unmarketable (Fullaway 1949). 'Smooth Cayenne' pineapple was shown resistant to natural infestation by these fruit fly species under commercial conditions (Flitters et al. 1953).

Regulatory agencies in different countries occasionally disagree about the host status of the same commodity for various reasons. The USDA Medfly Project Report (USDA 1930) described induced Mediterranean fruit fly infestations in cultivated Florida strawberries, *Fragaria virginiana* Duch. x *F. chiloensis* (L.). Strawberries also were listed as a Mediterranean fruit fly host in Europe by Thiem (1937), Bohm (1958), and Baas (1959), who reported finding Mediterranean fruit fly larvae in strawberries in Austria and Germany. Kobayashi & Fujimoto (1975) reared 26 Mediterranean fruit fly adults from infested strawberries collected at Kula, Maui (Hawaii). Armstrong et al. (1984) induced Mediterranean fruit fly infestations in Hawaii-grown strawberries in the field and recovered 125 and 66 Mediterranean fruit fly adults from unripe and ripe strawberries, respectively. Whereas the Japan Ministry of Agriculture, Forestry, and Fisheries regulates strawberries as a Mediterranean fruit fly host because of the field infestation information in the literature, USDA-APHIS-PPQ did not consider strawberries a host for Mediterranean fruit fly until after the work by Armstrong et al. (1984) because no Mediterranean fruit fly infestations were found during 30 years of inspections of strawberries from Spain (C. Amyx & G. Rohr, USDA-APHIS-PPQ, retired, personal communication). Conversely, Japan and Australia allow the importation of mature green bananas, *Musa acuminata* (Colla), from Central American countries where Mediterranean fruit fly and several *Anastrepha* species are known to occur. The U.S. requires quarantine treatment against Mediterranean fruit fly and oriental fruit fly for mature green bananas from Hawaii exported to the U.S. mainland although they are not a host for either fruit fly species at this stage of ripeness (Umeya & Yamamoto 1971, Armstrong 1983, Heather 1985).

Cowley et al. (1992) developed a protocol for determining the fruit fly host status of commodities. The protocol combines the determination of fruit fly presence in the fruit growing area, field collection of fruit, and both laboratory and field cage infestation studies. The host status protocol defines the experimental pathways needed to provide evidence for determining the host status of a fruit. Although the host status protocol developed by Cowley et al. (1992) may require some modification for different fruits or fruit fly species, it is the first published methodology that attempts to determine or verify the host status of fruits using logical and sequential approaches. Once the host status of the fruit is determined, the need for quarantine treatment can be decided (Cowley et al. 1992).

Moving host commodities through marketing channels without quarantine treatment is occasionally done under very controlled conditions. The host commodities are usually exported from pest free growing areas (Chapter 14), or during periods when the quarantine pest is absent. In Australia, some interstate markets, notably Victoria, accept fruit fly-host produce without quarantine treatment between May and August because fruit flies do not infest produce grown during these winter months. Although the infestation free growing period was based on proof of consistent freedom from infestation over many years and has worked satisfactorily in Australia, this quarantine strategy is not acceptable for export to overseas markets because of the required levels of security (Heather 1985).

A quarantine strategy that has not been fully exploited is the use of sorting methods. Sorting or culling to remove fruits that are or could be infested, such as overripe or damaged fruits, could be used as a quarantine treatment for some fruit fly hosts, especially poor hosts. However, regulatory agencies are uneasy about visual sorting. Objective scanning and detection equipment that guaranteed quarantine security would be needed for the implementation of any sorting method (Couey 1983).

Nonhost Cultivars

Different cultivars or hybrids of a single species can differ in host status. Flitters et al. (1953) showed that 'Smooth Cayenne' cultivar of pineapple was not a host for melon fly and oriental fruit fly. As new hybrids were developed from 'Smooth Cayenne,' each had to be tested for resistance to fruit fly infestation. Studies by Seo et al. (1973) on hybrid '59-443,' Armstrong et al. (1979) on hybrids 'D-10' and 'D-20,' and Armstrong & Vargas (1982) on cultivar '59-656' showed that these hybrids also were resistant to field populations of melon fly and oriental fruit fly. These four hybrids, and many new ones that were developed since, had

≥50% 'Smooth Cayenne' parentage. Based on the infestation studies with 'Smooth Cayenne' and the four hybrids, USDA-APHIS-PPQ determined that all hybrids with ≥50% 'Smooth Cayenne' parentage were resistant to infestation by melon fly and oriental fruit fly (USDA 1982). Other pineapple hybrids grown under similar conditions could be infested (Couey 1983). Each hybrid with ≤50% 'Smooth Cayenne' parentage would require testing for resistance.

Many species of *Citrus* are listed as good fruit fly hosts (Oakley 1950). However, Sproul (1976) found that green lemons, *Citrus limon* (L.) Burm. f., were resistant to infestation by Mediterranean fruit fly in Australia. Spitler et al. (1984) demonstrated that 'Eureka' and 'Lisbon' cultivars of commercial lemons from California were resistant to infestation by Mediterranean fruit fly. Nguyen & Fraser (1989) showed that 'Bearss' lemon and 'Persian' (actually 'Tahiti' cultivar according to Hennessey et al. [1992]) lime, *Citrus aurantifolia* (Christmann) Swingle, were resistant to infestation by Caribbean fruit fly, *Anastrepha suspensa* (Loew), in cage infestation tests although this fruit fly species reportedly attacks 11 other cultivars of citrus in Florida (Swanson & Baranowski 1972) (Japan presently imports lemons from Florida without quarantine treatment for Caribbean fruit fly [Nguyen & Fraser 1989]). Hennessey et al. (1992) demonstrated from large-scale fruit sampling over two years that 'Tahiti' lime was not a Caribbean fruit fly host under commercial conditions in Florida. Greany (1989) described three determinants involved with the mechanisms of resistance to fruit fly infestation by citrus fruits: the oviposition behavior of the fruit fly; the oil content of the peel and the presence of limonoids and naringin; and the softness of the peel. These determinants play a role in both cultivar resistance and nonhost stages of maturity.

Neilson (1967) and Pree (1977) showed that crab apples, *Malus toringoides* Hughes, were resistant to apple maggot fly, *Rhagoletis pomonella* (Walsh). Pree (1977) correlated the resistance to apple maggot infestation to the total phenol content of crab apples which prevented larval development in laboratory experiments.

Nonhost Stages of Maturity

Some commodities that are hosts for quarantine pests may not be infested at early stages of maturity. Harvesting a host at the noninfested stage of maturity can be used to avoid the need for quarantine treatment. There are a number of examples where a noninfested stage of maturity is used for this purpose.

Although ripe tomatoes are a host for Queensland fruit fly, *Bactrocera tryoni* (Froggatt), in Australia, green tomatoes are never infested. Therefore, tomatoes grown in Queensland can be shipped to Victoria if consignments are accompanied by a declaration stating that tomatoes were harvested in a hard green condition. Tomatoes can be ripened after harvest without jeopardizing the entry conditions (Heather 1985).

Banana can be infested by Mediterranean fruit fly, oriental fruit fly, Queensland fruit fly, and other fruit fly species when banana is mature (Armstrong 1983). However, Back & Pemberton (1916), Umeya & Yamamoto (1971), Armstrong (1983), and others showed that bananas were not infested by fruit flies at the mature green stage of ripeness, possibly because of either the tannin laden sap which prevents the female from ovipositing (Back & Pemberton 1916) or the exudation of latex from wounds, including oviposition sites (Armstrong 1983). Australia and Japan permit the importation of green bananas without quarantine treatment.

Partial Resistance to Infestation

Many commodities are relatively resistant to infection or infestation in the early stages of maturity. Senescent fruit are markedly more susceptible to fruit fly infestation than early-season fruit, probably because of a reduction in chemical components that are deterrents to oviposition or deleterious to development (Greany 1989). Seo et al. (1983) correlated resistance in unripe papayas to Mediterranean fruit fly, melon fly, and oriental fruit fly infestation to the release of linalool and benzyl isothiocyanate in response to damage. As papaya ripens, the ability to produce these compounds lessens and the fruit becomes more susceptible to infestation. Furthermore, although papayas have been recorded as a host for Caribbean fruit fly (Swanson & Baranowski 1972), Greany (1989) reported that it is rare for papayas to be infested in the field by this fruit fly species. Some cultivars of carambola, *Averrhoa carambola* L., are relatively resistant to Caribbean fruit fly infestation. Immature fruits are more resistant to oviposition than ripe fruits (Howard & Kenney 1987). Cultivars of melons, *Cucumis* species, are relatively resistant to melon fly infestation (Chelliah & Sambandam 1974a, b, Khandelwal & Nath 1978). Unfortunately, a "relatively resistant" commodity is still a host, albeit a poor one. Other steps must be taken to ensure quarantine security if no quarantine treatment is used. Grapefruit is a poor host for Caribbean fruit fly, especially at early harvest maturity (Calkins & Webb 1988). A quarantine protocol was developed to permit the shipment of early-season citrus and grapefruit from Florida to California without quarantine treatment. The protocol consists of harvesting grapefruit or other citrus during a specified time period (1 September through 20 December during 1991) from areas certified free of Caribbean fruit fly based on trapping or bait spray procedures (Voss 1990). This quarantine system uses the poor host status of grapefruit and other citrus to Caribbean fruit fly, combined with: harvest during a time of year when the fruits are most resistant to infestation (Calkins & Webb 1988); harvest during a cold period when Caribbean fruit fly is absent, in low numbers, or inactive; and harvest from areas certified as free of Caribbean fruit fly.

Mature avocados are considered good fruit fly hosts. However, research by Armstrong et al. (1983) and Armstrong (1991) found that 'Sharwil' avocado grown in Hawaii was not a host for Mediterranean fruit fly, melon fly, or oriental fruit fly when avocados were attached to the tree, probably because of fruit hardness and the ability of the fruit to isolate damage, including oviposition sites, with callus formation. Although Oi & Mau (1989) successfully infested with Mediterranean fruit fly and oriental fruit fly late-season 'Sharwil' avocados attached to the tree using severe artificial conditions, they did not find any natural fruit fly infestations. Armstrong (1991) found that 'Sharwil' avocados were not infested by oriental fruit fly under severe natural infestation pressure and reported that inspections of >114,000 individual 'Sharwil' avocados failed to produce a single fruit fly infestation. A quarantine protocol was developed whereby 'Sharwil' avocados could be exported from Hawaii to the U.S. mainland if unblemished mature green fruit with the stem attached were harvested from the tree and taken to a fly-proof packing house within 12 h of harvest (Anonymous 1990). The protocol, however, failed during the 1992 'Sharwil' avocado harvest for reasons still under investigation and the protocol was immediately rescinded (Anonymous 1992).

The 'Sharwil' avocado example underscores the fact that there are risks associated with using either nonhost stages of maturity or partial resistance to infestation as a basis for quarantine protocols. The subject commodity, environmental conditions, cultural practices, or the quarantine pest itself may change over time to defeat an inherent resistance to infestation. Continuous intensive field and/or packing house monitoring programs, fruit sampling regimens and/or infestibility tests may be required to maintain confidence in the quarantine security of quarantine protocols based on nonhost stages of maturity or partial resistance to infestation. While such quarantine protocols may include some risk and the options for overcoming the risks may be expensive, the loss or potential loss of quarantine fumigants (Armstrong 1992) requires that all available approaches to quarantine treatment or quarantine protocols be researched to maintain the flow of agricultural commodities through export markets.

Conclusions

For more than a decade, the trend in quarantine treatments research for consumable commodities has been to develop alternative treatments to toxic chemicals. The demand for chemical-free commodities requires new and creative quarantine strategies to maintain open marketing channels while preventing the spread of quarantine pests. Infestation-resistant or nonhost commodities, cultivars, stages of maturity, and growing periods offer alternatives to classical quarantine treatment technologies. Sorting or culling operations for commodities that are poor hosts also may be useful. These new strategies may require much more biological data from field and laboratory studies and more complex statistical evaluations to ensure quarantine security than were used for classical quarantine treatments. Inspection may become a key element to successful quarantine protocols using these strategies. The development of quarantine protocols using infestation resistant or nonhost commodities, cultivars, stages of maturity, and growing periods will require creativity and flexibility by researchers and regulators.

References

Anonymous. 1990. Department of Agriculture, Animal and Plant Health Inspection Service. Sharwil avocados from Hawaii, final rule. Federal Register 55: 38975-38978.

_____. 1992. Department of Agriculture, Animal and Plant Health Inspection Service. Sharwil avocados from Hawaii, interim rule. Federal Register 57: 31306-31307.

Armstrong, J. W. 1983. Infestation biology of three fruit fly (Diptera: Tephritidae) species on 'Brazilian,' 'Valery,' and 'William's' cultivars of banana in Hawaii. J. Econ. Entomol. 76: 539-543.

_____. 1985. "Pest Organism Response to Potential Quarantine treatments," in *Proceedings, Regional Conference on Plant Quarantine Support for Agricultural Development.* Pp. 25-31. Association of South East Asian Nations, Plant Quarantine Centre and Training Institute. Serdang, Selangor, Malaysia.

_____. 1991. 'Sharwil' avocado: Quarantine security against fruit fly infestation in Hawaii. J. Econ. Entomol. 84: 1308-1315.

_____. 1992. Fruit fly disinfestation strategies beyond methyl bromide. New Zealand J. Crop & Hortic. Sci. 20: 181-193.

Armstrong, J. W. & H. M. Couey. 1989. "Fumigation, Heat and Cold," in A. S. Robinson & G. Hooper, eds., *World Crop Pests, Vol. 3B, Fruit*

Flies. Their Biology, Natural Enemies and Control. Pp. 411-424. Amsterdam: Elsevier.

Armstrong, J. W. & R. I. Vargas. 1982. Resistance of pineapple variety '59-656' to field populations of oriental fruit flies and melon flies (Diptera: Tephritidae). J. Econ. Entomol. 75: 781-782.

Armstrong, J. W., J. D. Vriesenga & C.Y.L. Lee. 1979. Resistance of pineapple varieties D-10 and D-20 to field populations of oriental fruit flies and melon flies. J. Econ. Entomol. 72: 6-7.

Armstrong, J. W., W. C. Mitchell & G. J. Farias. 1983. Resistance of 'Sharwil' avocados at harvest maturity to infestation by three fruit fly species in Hawaii. J. Econ. Entomol. 76: 119-121.

Armstrong, J. W., E. L. Schneider, D. L. Garcia & H. M. Couey. 1984. Methyl bromide quarantine fumigation for strawberries infested with Mediterranean fruit fly (Diptera: Tephritidae). J. Econ. Entomol. 77: 680-682.

Armstrong, J. W., J. M. Harvey, D. L. Garcia, T. D. Menzes & S. A. Brown. 1988. Methyl bromide fumigation for control of oriental fruit fly (Diptera: Tephritidae) in California stone fruits. J. Econ. Entomol. 81: 1120-1123.

Baas, J. 1959. The Mediterranean fruit fly in Central Europe. Part 2. Hoefchen-Briefe Bayer Pflanzenschutz-Nactrichten 3: 113-140.

Back, E. A. & C. E. Pemberton. 1916. Banana as a host fruit of the Mediterranean fruit fly. J. Agric. Res. 5: 793-804.

Baker, A. C. 1939. The basis for treatment of products where fruitflies are involved as a condition for entry into the United States. USDA Circular 551.

Baker, R. T., J. M. Cowley, D. S. Harte & E. R. Frampton. 1990. Development of a maximum pest limit for fruit flies (Diptera: Tephritidae) in produce imported into New Zealand. J. Econ. Entomol. 83: 13-17.

Bohm, H. 1958. Zum Vorkommen der Mittelmeerfruchtfliege im Weiner Obstbaugebeit. Pflanzenschutzberichte 21: 129-158.

Calkins, C. O. & J. C. Webb. 1988. Temporal and seasonal differences in movement of Caribbean fruit fly larvae in grapefruit and the relationship to detection by acoustics. Fla. Entomologist 71: 409-416.

Chelliah, S. & C. N. Sambandam. 1974a. Mechanisms of resistance in *Cucumis callosus* (Rottl.) Cogn. to the fruit fly, *Dacus cucurbitae* Coq. (Diptera: Tephritidae). I. Non-preference. Indian J. Entomol. 36: 98-102.

_____. 1974b. Mechanisms of resistance in *Cucumis callosus* (Rottl.) Cogn. to the fruit fly, *Dacus cucurbitae* Coq. (Diptera: Tephritidae). II. Antibiosis. Indian J. Entomol. 36: 290-296.

Corey, F. M., Jr. 1985. "Survey on Mango Weevil, *Sternochetus mangifera*

F., in Central Visayas," in *Proceedings, Regional Conference on Plant Quarantine Support for Agricultural Development*. Pp. 63-66. Association of South East Asian Nations, Plant Quarantine Center and Training Institute. Serdang, Selangor, Malaysia.

Couey, H. M. 1983. Development of quarantine systems for host fruits of the Medfly. J. Hortic. Sci. 18: 45-47.

Cowley, J. M., R. T. Baker & D. S. Harte. 1992. Definition and determination of host status for multivoltine fruit fly (Diptera: Tephritidae) species. J. Econ. Entomol. 85: 312-317.

Flitters, N. E., F. Miyabara, S. Nakagawa & E. Dresner. 1953. The status of commercial pineapples as hosts of the oriental fruit fly in Hawaii. Special Report Ho-1, Fruit Fly Investigations in Hawaii. U.S. Dept. Agr., Entomology Res. Branch, Honolulu, Hawaii.

Fullaway, D. T. 1949. *Dacus dorsalis* Hendel in Hawaii. Proceedings, Hawaii Entomological Soc. 13: 351-355.

Greany, P. D. 1989. "Host Plant Resistance to Tephritids: an Under-exploited Control Strategy," in A. S. Robinson & G. Hooper, eds., *World Crop Pests, Vol. 3A, Fruit Flies. Their Biology, Natural Enemies and Control*. Pp. 353-362. Amsterdam: Elsevier.

Harte, D. S., R. T. Baker & J. M. Cowley. 1992. Relationship between preentry sample size for quarantine security and variability of estimates of fruit fly (Diptera: Tephritidae) disinfestation treatment efficacy. J. Econ. Entomol. 85: 1560-1565.

Heather, N. W. 1985. Alternatives to EDB fumigation as post-harvest treatment for fruit and vegetables. Queensland Agric. J. 3: 321-323.

Hennessey, M. K., R. M. Baranowski & J. L. Sharp. 1992. Absence of natural infestation of Caribbean fruit fly (Diptera: Tephritidae) from commercial Florida 'Tahiti' lime fruits. J. Econ. Entomol. 85: 1843-1845.

Howard, D. F. & P. Kenney. 1987. Infestation of carambolas by laboratory-reared Caribbean fruit flies (Diptera: Tephritidae): Effects of fruit ripeness and cultivar. J. Econ. Entomol. 80: 407-410.

Kahn, R. P. 1989. "Plant Protection and Quarantine," in *Biological Concepts, Vol. 1*. Boca Raton, Florida: CRC Press.

Khandelwal, R. C. & P. Nath. 1978. Inheritance of resistance to fruit flies in watermelon. Can. J. Genetics and Cytology 20: 31-34.

Kobayashi, R. & M. Fujimoto. 1975. U.S. Department of Agriculture, Agric. Res. Serv. Lab., semiannual report, June-Dec., 1975. Honolulu, Hawaii.

Landolt, P. J., D. L. Chambers & V. Chew. 1984. Alternative to the use of probit 9 mortality as a criterion for quarantine treatments of fruit fly (Diptera: Tephritidae)-infested fruit. J. Econ. Entomol. 77: 285-287.

Liquido, N. J., L. A. Shinoda & R. T. Cunningham. 1991. Host plants of the Mediterranean fruit fly, *Ceratitis capitata* (Wiedemann) (Diptera: Tephritidae): An annotated world review. Monographs Entomological Soc. America, Misc. Pub. 77.

Neilson, W.T.A. 1967. Development and mortality of the apple maggot, *Rhagoletis pomonella*, in crab apples. Can. Entomologist 99: 217-219.

Nguyen, R. & S. Fraser. 1989. Lack of suitability of commercial limes and lemons as hosts of *Anastrepha suspensa* (Diptera: Tephritidae). Fla. Entomologist 72: 718-720.

Oakley, R. G. 1950. "Fruit Flies (Tephritidae)," in *Manual of Foreign Plant Pests for Fruit Flies, Vol. 3.* Pp. 168-248. U.S. Department of Agriculture, Bureau of Entomology and Plant Quarantine, Division Foreign Plant Quarantine.

Oi, D. H. & R.F.L. Mau. 1989. Relationship of fruit ripeness to infestation in 'Sharwil' avocados by the Mediterranean fruit fly and the oriental fruit fly. J. Econ. Entomol. 70: 611-614.

Pree, D. J. 1977. Resistance to development of larvae of the apple maggot (Diptera: Tephritidae) in crab apples. J. Econ. Entomol. 70: 611-614.

Ruckelshaus, W. D. 1984. Ethylene dibromide, amendment of notice of intent to cancel registration of pesticide products containing ethylene dibromide. Federal Register 49: 14182-14185. U.S. Government Printing Office. Washington, D.C.

Seo, S. T., D. L. Chambers, C.Y.L. Lee, M. Komura, M. Fujimoto & D. Kamakahi. 1973. Resistance of pineapple variety 59-443 to field populations of oriental fruit flies and melon flies. J. Econ. Entomol. 66: 522-523.

Seo, S. T., C. S. Tang, S. Sanidad & T. H. Takenaka. 1983. Hawaiian fruit flies (Diptera: Tephritidae): variation of index of infestation with benzyl isothiocyanate concentration and color of maturing papayas. J. Econ. Entomol. 76: 535-538.

Spitler, G. H., J. W. Armstrong & H. M. Couey. 1984. Mediterranean fruit fly (Diptera: Tephritidae) host status of commercial lemon. J. Econ. Entomol. 77: 1441-1444.

Sproul, A. N. 1976. Green lemons safe from fruit fly. J. Agr. Western Australia 17: 32.

Swanson, R. W. & R. M. Baranowski. 1972. Host range and infestation by the Caribbean fruit fly, *Anastrepha suspensa* (Diptera: Tephritidae), in south Florida. Proceedings, Fla. State Hortic. Soc. 85: 271-274.

Thiem, H. 1937. Die Mittelmeerfruchtfliege. Flugblatt No. 151: 1-6.

Umeya, K. & H. Yamamoto. 1971. Studies on the possible attack of the Mediterranean fruit fly (*Ceratitis capitata* [Wiedemann]) on the green bananas. Research Bul. Plant Protection Service (Japan) 9: 6-17.

USDA. 1930. USDA Mediterranean Fruit Fly Project Report of the Biological Research Division. April 1929 to Feb. 1930. Cage

Experiments, Orlando, Florida.

_____. 1982. Hawaiian fruits and vegetables, M318.13. Plant Protection and Quarantine Programs Port of Entry Manual. Animal and Plant Health Inspection Service, Hyattsville, Maryland.

_____. 1983. Host list: Mediterranean fruit fly, *Ceratitis capitata* (Wiedemann). Biological Assessment Support Staff, Plant Protection and Quarantine, Animal and Plant Health Inspection Service, Hyattsville, Maryland.

Vargas, R. I. & T. Nishida. 1985a. Life history and demographic parameters of *Dacus latifrons* (Hendel) (Diptera: Tephritidae). J. Econ. Entomol. 78: 1242-1244.

_____. 1985b. Survey for *Dacus latifrons* (Diptera: Tephritidae). J. Econ. Entomol. 78: 1311-1314.

Voss, H. J. 1990. Permit QC 222, master permit for the shipment of citrus fruit from the state of Florida to California, amended Sept. 20, 1990. California Department of Food and Agriculture, Sacramento. 14 pp.

Wester, P. J. 1920. The mango. Philippine Bureau Agr. Bul. 18.

14

Pest Free Areas

Connie Riherd, Ru Nguyen, and James R. Brazzel

The pest free area concept is a program that has been used for almost a decade to certify fruit for export to meet quarantine restrictions. The concept mainly is utilized to certify areas free from economically important fruit flies. The first program to certify pest free areas was one used for citrus production areas in Texas certified free of Mexican fruit fly, *Anastrepha ludens* (Loew). Similar programs are in progress elsewhere for other pests.

The pest free areas discussed herein are ones in which an organized program has been developed and designed to detect the presence of the pest in numbers great enough to constitute a pest risk from the regulatory standpoint. The program includes suppression measures which may be utilized in a prophylactic mode to prevent population development plus measures to reduce populations. The procedures ensure that the area meets protocol standards for certification. In effect, the present programs are pest management programs designed to keep pest populations below pest risk levels. This allows certification in the field of the product based on regulatory requirements.

The basic tool for management of pest free areas is a trap array established in the fields that will be certified. Programs to date are primarily in areas subject to infestation by *Anastrepha* species. The McPhail trap is the primary detection device used for certification survey of *Anastrepha* species (Fig. 14.1). The survey data generated from the trap arrays are used to activate various options available to regulatory officials such as, (1) continue certification, (2) withdraw certification, or (3) initiate suppression measures which will bring the area back into compliance. The criteria for action on these options are provided in the

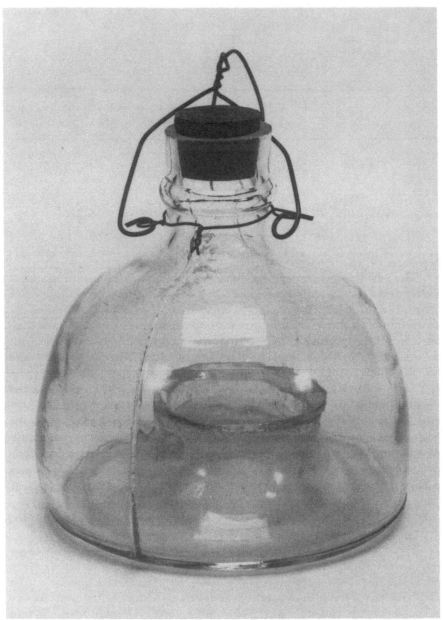

FIGURE 14.1. McPhail trap used to detect Caribbean fruit fly.

program protocol developed by persons concerned with the program, including producers.

The fate of these programs depends upon the efficiency of the trap survey. Data must reliably indicate the pest situation in the field by comparing the number of adults caught in traps with the number of larvae detected by cutting fruit. Trap data are validated as a basis for protocol development and regulatory action by this procedure.

In some programs, pest suppression measures are used to ensure protocol requirements. Occasionally, sterile fly releases or malathion bait sprays are used to prevent population increases to levels which would result in the loss of pest free certification status. The bait spray also may be used to reduce populations to levels which would be within the limits for certification.

Mexican Fruit Fly in Texas

Before 1981, citrus fruits from areas infested with Mexican fruit fly were fumigated after harvest for export purposes. The preferred fumigant was ethylene dibromide (EDB). Based on concerns by regulatory agencies and affected industries that the use of EDB might eventually be limited, an alternative method was investigated to certify citrus fruit free of Mexican fruit fly in the Rio Grande Valley of Texas. This method involved the certification of specific production areas as being maintained free of Mexican fruit fly. Pest risk was considered negligible if the regulatory measures which were in place to suppress the infestation were modified to, (1) require trapping surveillance at an increased trap density (five traps per square mile of citrus), (2) limit nonfumigated fruit movement from these areas to specific months (prior to January 1), and (3) suspend certification if five or more flies were found (B. Granberry, personal communication). The rationale for certification as fly free was based on scientific and historical data and on experience. Low numbers of Mexican fruit fly were detected from July to April indicating that the population developed each year either from new introductions from Mexico or from very low level or nondetectable populations which survived the low availability of hosts in the summer. Regardless, the fly population at the beginning of the period when citrus was susceptible to attack (after color break of fruit) was very low or nonexistent. Populations of Mexican fruit fly were further suppressed by the release of sterile flies. Also, evidence suggested that the Rio Grande Valley was the northern limit of Mexican fruit fly. Availability of hosts, especially during June - August, indicated that the environment exerted a strong suppression effect on fly population (B. Granberry, personal communication).

Other citrus producing states expressed concern over the implementation of this new certification procedure, which went into effect on 2 November 1981. Therefore, the fly free protocol was modified to include using McPhail traps, cutting grapefruit to check for larvae (30,000 fruits cut prior to implementation and 30,000 cut over the first 30 d thereafter to validate the trap data), and withdrawing certification if, (1) one larva or one gravid female was found, (2) one additional male or nongravid female was found in a radius of 1.5 miles of the original find, or (3) greater than three flies were found in traps.

Research continued over the next two to three years to further refine the management system and to validate the reliability of the trap survey as a means for monitoring the program. Refinement was needed to convince other citrus producing states and countries that the approach provided quarantine security. The research was concentrated in three areas that were to, (1) validate the McPhail trap as a reliable survey tool by cutting 10 - 20 thousand fruits over a two-year period in the areas with traps arrayed at five per square mile (the presence of the pest in an area could be detected at least one generation earlier with adult trapping than by fruit cutting for larval detection), (2) evaluate sterile Mexican fruit fly for population suppression (sterile flies used in these trials exerted as much as 75% suppression on native populations), and (3) evaluate malathion bait spray for control of Mexican fruit fly. (A 10 square mile aerially applied bait spray test performed in Mexico against Mexican fruit fly indicated treatments at 10 d intervals over a four- to five-month period eradicated the population.) (J. R. Brazzel, unpublished data.)

A protocol was developed based upon year-round dispersal of sterile flies, use of bait spray as needed, and use of McPhail traps for survey to certify citrus as fly free and is being used. To date no Mexican fruit flies have been detected in the certified citrus.

Caribbean Fruit Fly in Florida

Caribbean fruit fly, *Anastrepha suspensa* (Loew), was reported in Florida in 1965 in large numbers (Weems 1966). No attempts to eradicate this pest were made since the fly had been studied for some 30 years in Puerto Rico, where it was observed to generally ignore citrus (Anonymous 1966, Weems 1966). However, citrus is a host of Caribbean fruit fly in Florida, and for this reason, must be treated to meet the quarantine requirements of Arizona, California, Hawaii, Texas, Japan, and

Bermuda. The preferred hosts of Caribbean fruit fly include rose apple, *Syzygium jambos* (L.) Alston, cattley guava, *Psidium littorale* Raddi var. *longipes* (O. Berg) Fosb., Surinam cherry, *Eugenia uniflora* L., common guava, *Psidium guajava* L., and loquat, *Eriobotrya japonica* (Lindl.) (Swanson & Baranowski 1972).

The shipment of citrus fruit from pest free areas in Florida began during the 1982-83 season when California and Texas accepted grapefruit from production areas certified free of Caribbean fruit fly. To meet pest free area protocol requirements, the production area had to be a minimum of 120 hectares (300 acres) and located at least 4.8 kilometers (3 miles) from any residential areas having numerous Caribbean fruit fly preferred hosts. Since citrus fruit from Florida could be fumigated with EDB for shipment to California and Texas, only Fellsmere Farms in Indian River County, and Cow Creek and Blue Goose in St. Lucie County, encompassing 890 hectares (2,200 acres), participated in the program. The program was terminated in September 1984 due to the detection of citrus canker in Florida (Griffith 1986).

The Environmental Protection Agency (EPA) announced that the use of EDB for postharvest quarantine treatment would be terminated in 1986 (Anonymous 1984). Intensive field research conducted in Indian River County during 1984-87 and experiences with Mexican fruit fly in Texas were used to develop pest free areas for shipping citrus fruit to Japan (Nguyen et al. 1992). A proposed protocol for shipping fresh grapefruit from pest free areas was submitted to Japan in 1986 based on data which indicated that, (1) grapefruit and orange are minor hosts of Caribbean fruit fly, (2) most Caribbean fruit fly are concentrated in urban areas where many major hosts are sustained year round, (3) extremely low numbers of Caribbean fruit fly move from urban areas to commercial citrus groves, and (4) low numbers of Caribbean fruit fly occur during the shipping season in winter months (January-mid April). This protocol was a modification of the original domestic program and used solely trapping to certify a pest free product. The production areas (called "designated areas") had to be a minimum of 120 hectares (300 acres) and at least 4.8 kilometers (3 miles) from preferred hosts which were found mostly in urban areas. The designated area and its surrounding buffer zone of 2.4 kilometers (1.5 miles) could not contain preferred host plants. McPhail traps were used to monitor Caribbean fruit fly in the designated area and its buffer zone at the rate of 15 traps per square mile. Grapefruit could be shipped without fumigation as long as no flies were detected.

The proposal was approved in August 1986. Grapefruit from Indian River and St. Lucie counties was shipped to Japan in October.

Many citrus groves in Florida were either too small or too close to urban areas to meet the protocol. Modifications of the protocol were developed to allow for greater use. Aerial bait sprays were incorporated into the Japanese protocol during the 1987-88 season. Minimum size of the designated areas was reduced to 16 hectares (40 acres) with a 0.5 mile buffer zone. The designated area and 92 meters (300 feet) surrounding barrier environs had to be trapped 30 d before harvest and bait sprayed at 7 - 10 d intervals through the harvest season.

The program developed further as new information became available. Less restrictive certification criteria for early-season grapefruit was based on its resistance to Caribbean fruit fly infestation (Greany et al. 1983, Greany et al. 1987) and field tests where substantial fruit cutting data collected during November and December showed no infestation by Caribbean fruit fly.

The current protocol for certifying grapefruit and orange for export from areas maintained free of Caribbean fruit fly contains two distinct certification routes. The first is based on trapping, where the production area must be >4.8 kilometers (3 miles) from residential or other areas where numerous preferred fruiting host plants are located. The designated area must be 120 hectares (300 acres) in size and surrounded by a buffer zone of 0.8 kilometer (0.5 miles) which should not contain preferred host plants. If host plants occur in the buffer, they must be treated with bait spray, 178 ml of 91% malathion, and 710 ml of protein hydrolyzate bait per hectare on a weekly basis. The area must be trapped with a density of 15 McPhail traps per square mile (2.56 square kilometers) and traps must be serviced weekly. If two flies are found within 2.4 kilometers (1.5 miles) of each other during a 30 d period, an area of 2.4 kilometers (1.5 miles) around the detection is immediately withdrawn from the program. One trapped fly does not initiate action since it is not considered to be an indication that an infestation exists in the designated area. For early-season grapefruit (1 August - 20 December), the detection of three flies initiates withdrawal of certification for the area. Areas can be recertified if a 0.4 kilometer (0.25 mile) area around the trap catch is treated with bait spray at 7 - 10 d intervals 30 d prior to and during the harvesting period and traps indicate a fly-free condition. There is no 30 d suspension period for early-season grapefruit. Also, trap densities are reduced for early-season grapefruit requiring only two traps per square mile (2.56 square kilometers) in the production area and five per square mile (2.56 square kilometers) in the buffer.

The second certification procedure is based on the application of bait sprays. The minimum size of the production area must be 16 hectares (40 acres) with an additional buffer zone of 92 meters (300 feet). The buffer cannot contain any preferred host plants. The area must be at least 0.8 kilometers (0.5 mile) from residential areas where numerous fruiting preferred host plants occur. Host plants within this area must be treated with bait spray on a weekly basis. Traps must be established in the area (and buffer) at the density of 15 traps per square mile (2.56 square kilometers) with a minimum of four traps required. Aerial bait sprays consisting of a mixture of 178 ml of 91% malathion and 710 ml of protein hydrolyzate bait per hectare must be applied to the area and the buffer zone of 92 meters (300 feet) beginning 7 d before to harvest and throughout the harvest period at 7 - 10 d intervals. Early-season grapefruit can be certified under this procedure without the requirement of the production area being 0.8 kilometer from residential areas or areas with numerous preferred hosts. If a fly is found in these areas after bait spray begins, the areas are withdrawn from the program.

Important aspects of the program must be monitored closely to guarantee the integrity of the certification procedure. Records are needed to maintain the identity of fruit from the point of harvesting throughout the packing and shipping procedure. The fruit harvest boxes are labelled with an identification number which designates the production area from which the fruit was picked. This number follows that fruit to the point of destination. Harvesting and packing house procedures are closely monitored by regulatory officials. Should a fly infestation be found during the import inspection, only fruit from that designated area is rejected. This has not occurred in any fruit certified in this manner since the program was implemented. Another important aspect is the survey for preferred hosts around the production areas which are conducted by regulatory officials during the off-season summer months. Plants are tagged for removal by the grower and trapping of the area does not commence until removal is verified. All phases of the program including host survey, trapping, maintaining fruit identity, bait spray treatments, packing, and loading are either conducted by or closely monitored by regulatory officials. The cost of the certification program is paid by the growers who are assessed a per acre charge each month of participation.

Participation in the program continues to increase. Nineteen certified counties were involved in 1993-1994 harvesting season (Fig. 14.2). Participators continue to apply to have new areas certified as fly free. Future plans to expand and further develop the program may include the incorporation of sterile fly release (T. C. Holler & E. J. Nickerson, unpublished data) and the release of a parasitic wasp, *Dichasmimorpha longicaudata* (Ashmead) (Thompson 1989). These techniques would be used to suppress fly populations in urban areas where Caribbean fruit fly populations are high due to preferred host concentrations. Also, the use of gibberellic acid which delays the ripening of the peel and extends resistance to infestation (Greany et al. 1987) might play a role for the certification of fruit in production areas where preferred host removal is not practical.

Participating Counties
(in shaded area)

Polk	Highlands
Hardee	Okeechobee
De Soto	Brevard
Charlotte	Indian River
Lee	St. Lucie
Collier	Martin
Hendry	Palm Beach
Glade	Sarasota
Volusia	Manatee
Osceola	

FIGURE 14.2. Florida map showing counties participating in pest free areas for Caribbean fruit fly, 1993-94.

Pest Free Areas in Other Countries

The pest free area concept to satisfy quarantine restrictions is gaining acceptance by affected industries and regulatory agencies in both producing and receiving states and countries. The United States (U.S.) is accepting fruit from Mexico, Brazil, Ecuador, and other countries certified in this manner based on approval by the U.S. Department of Agriculture, Animal and Plant Health Inspection Service, Plant Protection and Quarantine (USDA-APHIS-PPQ).

There are several common factors in certifying host fruit as originating from an area free of fruit flies including some form of geographic separation of the production area from infested areas, a trapping system to verify the absence of fly infestation, and the maintenance of the identity of fruit harvested from certified areas to prevent mixing with fruit from noncertified areas. Criteria vary depending on the fruit fly species, the host, and the geographic location. Certification criteria may include the incorporation of sterile fly releases, prophylactic bait sprays, preferred or alternate host removal, the utilization of trap crops, and the inspection for larvae through fruit cutting or incubation procedures.

The basic criteria for establishing fruit fly free areas can be expanded to other pests of quarantine significance. The USDA-APHIS-PPQ has established some guidelines for countries desiring to export products to the U.S. under the terms of "free area regulations" (M. F. Kirby, personal communication). Candidate pests are ones of quarantine significance for which product inspection for the pest is not an acceptable method for providing quarantine security and for which a quarantine treatment is either not desirable, acceptable, or available to the importing country. Also, it must be technologically, politically, and financially feasible to establish or maintain specific areas of the country of origin as free from the target pest. The exporting country must ensure support and cooperation of industry groups (producers and exporters) who will receive benefit, financially support, and provide daily operational support for the program. Also, they must ensure support and cooperation of government entities (primarily the National Plant Protection Services) who will direct and enforce program technical requirements and carry out negotiations with the USDA-APHIS-PPQ. After criteria are satisfied, a work plan must be developed (Table 14.1). The work plan must contain proven survey technology and quarantine regulations to prevent introduction of the target pest into the free area and include full details of the action which will be taken in the event that the target pest is detected in the free area (i.e., cancellation of free area status, eradication measures, and surveys to demonstrate recovery of free status). Procedures for marking and certifying the product which is to be exported from the free area and for safeguarding it to prevent infestation after leaving the free area must also be included in the work plan. The USDA-APHIS-PPQ requires that the work plan be implemented for a minimum of one year and the data show nonoccurrence of the target pest in the proposed free area. Approval by the USDA-APHIS-PPQ will be based on the review of the work plan and data, an on site visit, and publication in the Federal Register to formally establish the existence of the approved, pest free area. The continued acceptance of products exported from the approved pest free area depends on continued satisfactory execution of the approved work plan based on site review.

TABLE 14.1 General requirements for pest free areas

I.   Development of a work plan.
     A. Survey technology.
     B. Quarantine regulations to prevent pest introduction.
     C. Action to be taken if pest is found.
     D. Procedures for marking or identifying the product, and for certifying and safeguarding the product.
II. Implementation of work plan for one year with non-occurrence of the target pest.
III. Review of the work plan and data by the USDA-APHIS-PPQ.
IV. Site visit by the USDA-APHIS-PPQ.
V.   Publication in the Federal Register.
VI. Continued satisfactory execution of the approved work plan based on site review.

Expenses for on site review are reimbursable to the USDA-APHIS-PPQ. Some examples of products certified for import into the U.S. from approved pest free areas include apples, grapefruit, oranges, peaches, and tangerines from the Sonora, Mexico, municipalities of Altar, Atil, Caborca, Carbo, Hermosillo, Pitiquito, Puerto Penasco, and San Miguel. The areas are approved as being free of Mediterranean fruit fly, *Ceratitis capitata* (Wiedemann), Mexican fruit fly, *Anastrepha serpentina* (Wiedemann), West Indian fruit fly, *Anastrepha obliqua* (Macquart), and *Anastrepha fraterculus* (Wiedemann). Honeydew melons from areas in both Brazil and Ecuador also are approved as being free of *Anastrepha grandis* (Macquart) (Anonymous 1992).

Conclusions

Current pest free area programs are proven effective mechanisms for meeting quarantine restrictions. A variety of techniques has been incorporated to develop these areas and to maintain pest free status. The development of new pest free areas may be complex and will certainly vary depending on the target pest and the host product. The criteria used must be biologically justified with acceptable supporting data, and will undoubtedly be subject to change as new techniques are developed or as new information becomes available. The continued development of such techniques and the increasing demand for high quality fresh fruit and vegetables and other host products will ensure the continued expansion of the pest free area concept.

References

Anonymous. 1966. Researchers seek key to Caribfly. Fla. Department of Agr. & Consumer Services, Division of Plant Industry News Bul. 8: 1-4.

_____. 1984. EPA proposes residue levels for EDB in citrus. EPA Environmental News. March 1984. Washington, D.C.

_____. 1992. Animal and Plant Health Inspection Service. Plant protection and quarantine treatment manual. U.S. Government Printing Office. Washington, D.C.

Greany, P. D., S. C. Styer, P. L. Davis, P. E. Shaw & D. L. Chambers. 1983. Biochemical resistance of citrus to fruit flies. Demonstration and elucidation of resistance to the Caribbean fruit fly, *Anastrepha suspensa*. Entomologia Experimentalis et Applicada 34: 40-50.

Greany, P. D., R. E. McDonald, P. E. Shaw, W. J. Schroeder, D. F. Howard, T. T. Hutton, P. L. Davis & G. K. Rasmussen. 1987. Use of gibberellic acid to reduce grapefruit susceptibility to attack by the Caribbean fruit fly, *Anastrepha suspensa* (Diptera: Tephritidae). Tropical Sci. 27: 261-270.

Griffith, R. J. 1986. Citrus canker eradication program. Fla. Dept. of Agr. & Consumer Services, Division of Plant Industry, 36th Biennial Report. Pp. 71-74.

Nguyen, R., C. Poucher & J. R. Brazzel. 1992. Seasonal occurrence of *Anastrepha suspensa* (Diptera: Tephritidae) in Indian River County, Florida 1984-87. J. Econ. Entomol. 85: 813-820.

Swanson, R. W. & R. M. Baranowski. 1972. Host range and infestation by the Caribbean fruit fly, *Anastrepha suspensa* (Diptera: Tephritidae), in south Florida. Proceedings, Fla. State Hortic. Soc. 85: 271-274.

Thompson, C. R. 1989. *Dichasmimorpha longicaudata* (Ashmead) (Hymenoptera: Braconidae), biological control agent for the Caribbean fruit fly. Fla. Department of Agr. & Consumer Services, Division of Plant Industry, Entomol. Circular 325: 2 pp.

Weems, H. V., Jr. 1966. The Caribbean fruit fly in Florida. Proceedings, Fla. State Hortic. Soc. 79: 401-405.

15

Systems Approaches to Achieving Quarantine Security

Eric B. Jang and Harold R. Moffitt

The development of quarantine treatments for use against pests that infest food plants arose from the need to protect agricultural interests without prohibiting commerce. Quarantine procedures are necessary to ensure that exotic pests do not enter a geographic location where they do not currently exist. Although the need for establishing quarantine security is practically universal among nations, guidelines for exactly what constitutes quarantine security are not universal. The majority of the quarantine treatments used to ensure that commodities are free from exotic pests have been direct treatments (e.g., heat, cold, fumigants) that assume a high (and frequently unknown) level of pest infestation in the commodity. These treatments are perceived as providing quarantine security if they meet an established mortality level based on laboratory and field efficacy tests. For example, treatments for fruits infested with tephritid fruit flies are usually required to demonstrate mortality of at least 99.9968% (probit 9) or no more than 32 live insects per million of treated insects (Baker 1939). However, many commodities that have been reported as hosts for various pests are at best poor hosts and do not normally sustain high infestation levels in the fruit. In other cases, the host commodity is not infested if certain operational procedures are met (e.g., maintaining fly free areas, harvest at specified maturity, pest-host phenology). In such situations, integrating biological information about the pest, knowledge of the host-pest relationship, and incorporating specific operational factors leading to and including the grading, sorting, and packing of the commodity may be sufficient to meet quarantine security. In some cases, this information alone may provide the framework for certification without the need for a direct treatment. The integration of such biological and operational factors into a viable system of procedures that will itself meet quarantine security is the topic of this chapter. In it are discussed the concepts that comprise the systems approach, definition of a system, and differentiation of systems approaches from direct quarantine treatments or multiple treatments.

The systems approach to quarantine treatments is a concept that has evolved in response to the need to consider the various preharvest and postharvest biological factors that can influence the level of infestation leading to and including the sorting, packing, shipping, and marketing of commodities. To understand the systems approach and how it differs from other methods used to certify that a commodity is free from a pest, one must define the terms that will be used in this chapter and clarify how the systems approach differs from methods described in other chapters included in this book. The systems approach can be defined as the integration of those preharvest and postharvest practices used in production, harvest, packing, and distribution of a commodity which cumulatively meet the requirements for quarantine security. Systems approaches integrate biological, physical, and operational factors that can affect the incidence, viability, and reproductive potential of a pest into a system of practices and procedures that together provide quarantine security.

Comparison of Systems Approaches
with Other Quarantine Methods

Systems approaches (hereafter referred to as systems) differ from other methods such as direct treatments in that they integrate biological and operational factors to meet quarantine requirements. Unlike commodities that are considered nonhosts, systems recognize that the commodity in question is a host, the level of infestation in the host being the key component in the design of the overall system. Systems rely on knowledge of the infestation level of the host and measure the impact of the various operational procedures on removing infested hosts, thereby reducing the risk that infested fruits will be shipped. Systems can be differentiated from single quarantine treatments such as fumigation, heat, cold, irradiation, and atmospheric modification (modified or controlled) in which the treatment alone is sufficient to ensure quarantine security. Systems should also be separated from the concept of combination and multiple treatments. Combination and multiple treatments are comprised of two or more direct treatments each of which, if used alone, would not be sufficient to achieve quarantine security (Chapter 16). Each part of combination and multiple treatments must have a calculated level of security that, when combined, will ensure adequate security. However, a direct treatment can be a systems component if the treatment alone will not provide adequate security, but in combination with other (operational) parts of the system will provide quarantine security. This working definition of how systems are used in providing quarantine security can be better understood within the context of a larger discussion of how specific components of systems work and relate to each other.

Components of Systems

Because systems attempt to integrate biological, physical, and operational factors into coherent practices, the repertoire of techniques which could be considered as part of systems is large and diverse. For the purpose of this discussion we have chosen to identify several key components that can be included in systems. However, it must be emphasized that within each component are several practices that can be combined in different ways with other components to develop systems. A generalized flow chart showing the components of a system is shown in Figure 15.1.

Production Component

Systems should integrate all available production practices that reduce or limit the pest population in the field. Thus, integrated pest management (IPM) practices are frequently part of systems. Normal IPM practices usually rely on inherent knowledge of the biological status of the pest with regard to the host in question. This information is the foundation from which accurate assessments are made regarding the use of other systems components. Pest surveys and sampling are routinely used in IPM programs to define economic thresholds or economic injury levels of the crop being grown. Predictive models are frequently helpful to determine initial and ongoing pest population levels. Results of surveys and samples could trigger the use of pest management practices to reduce population levels in the field.

Shipments of Florida-grown grapefruit to California include a field monitoring program and bait sprays as part of the system to ensure that areas are free from Caribbean fruit fly, *Anastrepha suspensa* (Loew) (Chapter 14). Apple growers in the Pacific Northwest area of the United States (U.S.) monitor codling moth, *Cydia pomonella* (L.), field population levels and integrate this information to determine an appropriate use of selective chemical sprays when a certain economic threshold is reached.

PRODUCTION

no

⬇   ⟨ᶠPIT ᶳ⟩

yes

[Inputs] Population surveys,
Predictive models, control measures
based on pest incidence thresholds

PREHARVEST

no

⬇   ⟨ᶠPIT ᶳ⟩

yes

[Inputs] Infestation biology of pest,
Commodity/Pest phenology, Pest
incidence thresholds

POSTHARVEST

no

⬇   ⟨ᶠPIT ᶳ⟩

yes

[Inputs] Procedures to reduce
remaining infestation (direct
treatment), Culling, sorting,
packing, Prevention of
reinfestation

INSPECTION /
CERTIFICATION

no

⬇   ⟨ᶠPIT ᶳ⟩

yes

[Inputs] Usually based on security
requirements of the importing country

SHIPPING /
DISTRIBUTION

[Inputs] Further reduction in pest
"risk" through marketing channels

FIGURE 15.1.   Components of the systems approach to achieving quarantine security. Movement to a subsequent level is dependent on meeting pest incidence thresholds (PIT) or appropriate control measures must be initiated.

Systems should rely on production estimates of field populations at harvest. More importantly, there is a need to improve correlations between field populations and resulting in-field infestation levels at harvest that impact systems.

Preharvest Component

The preharvest component of systems uses information on the pest's interactions with the host commodity at or about the time of harvest to further reduce the risk that a commodity entering the postharvest chain will be infested. Accurate knowledge of the infestation biology of the pest and the commodity at the time of harvest along with knowledge of the fruiting phenology and ripening of the host commodity may reveal specific seasonal periods where infestation would not occur or would be light. Components showing pest-host asynchrony in certain geographic locations, the use of specific fruit selection criteria such as ripeness, hardness, and color, or specific harvesting techniques meant to reduce or alleviate infestation, all have been included as parts of proposed systems for different commodities. Progress in these areas has been made in the use of this component in systems. Emphasis on better knowledge of the infestation levels and biologies of quarantine pests should continue in the development of systems. Yokoyama et al. (1992) suggested that California-grown nectarines shipped to New Zealand would not be infested with walnut husk fly, *Rhagoletis completa* Cresson, due to early season asynchrony between the emergence of the pest in walnuts and the presence of early-season nectarine fruits in the San Joaquin Valley. Curtis et al. (1991) surveyed codling moth levels in the field as well as in culled and packed nectarines to determine the incidence of codling moth infestation at harvest. Couey & Hayes (1986) recommended a fruit selection procedure based on fruit color as part of the double hot water dip system for fruit flies in Hawaii-grown papayas based on the work of Seo et al. (1982) who reported that mature green fruits were seldom infested. This was highly correlated with levels of the chemical benzyl isothiocyanate in green fruit (Seo & Tang 1982, Seo et al. 1983). Armstrong et al. (1983) showed that 'Sharwil' avocados while on the tree were not normally a host of fruit flies in Hawaii.

Postharvest Component

Historically, most quarantine treatments have focused on the postharvest component of systems as the starting point in the development of direct treatments. The lack of specific knowledge of how the production and preharvest components impact field infestation levels has hindered the development of true systems, leading to treatments where maximum infestation levels were assumed. Central to the success of any systems to quarantine treatments is the determination of an initial threshold or tolerance level of infested fruit upon arrival at the packing house. This assessment is the key to determining the effectiveness of any subsequent operational procedures (including direct treatments) aimed at removing remaining infested fruit. The initial tolerance level upon arrival at the packing house should be low enough to ensure that subsequent procedures will remove remaining infested fruit sufficient to provide quarantine security. Specific packing house procedures can include but are not limited to culling, grading, sorting, and packing. Knight & Moffitt (1991) found that several factors including the degree of injury in the fruit upon arrival at the packing house, volume of apples sorted per worker second, and the appearance and size of injury all affected the culling procedure. A mean 84% of codling moth-injured fruits were removed by culling, with initial injury levels of 0.2 - 35% to the fruit upon arrival at the packing house. Direct treatments used as a component of systems would be those not providing quarantine security when used alone. However, when used in conjunction with production and/or postharvest components, these direct treatments contribute to systems that provide quarantine security. Packaging of the commodity to ensure that reinfestation does not occur could include methods such as shrink wrapping (Jang 1990) or the use of specially designed boxes to prevent possible reinfestation.

Final Inspection and Certification

Final inspection of the packed product is necessary to ensure that the product to be shipped has gone through the proper inspections/treatments. Certification, normally carried out by the importing country, can be included as part of the final inspection. Although most of the final inspection certification is currently done prior to shipment (e.g., security was determined prior to shipment), certification at destination is another option that needs to be considered as part of systems to quarantine treatments. Couey et al. (1984) developed a system for Hawaii-grown papayas that included an in-transit cold treatment that would be certified at destination.

Marketing and Distribution

Perhaps the least studied area in systems is the marketing and distribution of the commodity and its impact on pest risk assessment. Most commodities are divided into smaller units upon arrival at the destination and then further divided as they enter retail marketing channels. The impact of such further divisions, and the determination of exactly what makes up a unit for the purpose of risk assessment at destination, has not been well defined. Lacking such a consensus, the ability to realistically factor this information into systems will be limited. Including biological factors into systems requires knowledge of the postharvest biology of the pest in question to determine the probability of a mating pair arriving at the same destination or location in the importing country. Vail et al. (1993) studied the biological factors as well as postharvest packaging and distribution practices that would normally be a part of the marketing system in Japan in assessing quarantine security against codling moth inside in-shell walnuts shipped from California. Better knowledge of such systems would serve to further identify how operational factors and biological factors could be integrated into more comprehensive systems.

There are many ways in which each of the above components could be integrated to form a system providing quarantine security. The ability of any one of the components to impact the pest population and subsequent infestation will usually be dependent on that component alone. However, the success of that component will impact the success of subsequent components that make up a system. At each step, assessment of the pest incidence threshold allows for the determination of how subsequent components of a system will impact infestation levels. For example, the success of a system that uses culling, sorting, and packing of a commodity in lieu of a direct treatment would be dependent on a low tolerance level for the pest upon arrival at the packing house (i.e., production and preharvest methods combining to reduce initial infestation) as well as a calculated efficiency in removing damaged fruit on the packing line.

Systems to Achieve Quarantine Security

Systems to achieve quarantine security have been used in the past although perhaps not widely recognized. Specific examples of systems and how systems are integrated to provide quarantine security are discussed for apple, cherry, papaya, in-shell walnut, and watermelon.

Export of Pacific Northwest Apples and Cherries

The Pacific Northwest currently ships apples and cherries to several Pacific Rim countries using systems to ensure that these fruits are not infested with codling moth. Apples are a preferred host of codling moth while cherries are a rare or sporadic host in the U.S. (Moffitt 1990). The system used for apples includes control of codling moth in the orchard (production); inspection of harvested fruit on arrival at the packing house with rejection of those lots not meeting the packing house's established standards; grading, sorting, and packing emphasizing removal of codling moth injured fruit (postharvest); and a final inspection and certification that quarantine security has been achieved. In his study of a system used for export over a two year period, Moffitt (1990) found that only 10 out of 171,448 culled apples were infested with codling moth. None of the 501,537 apples examined from boxes packed for export were infested. Also, data from the Washington State Department of Agriculture showed that only 33 of 41,397,020 apples inspected for export over a five year period were infested. These data strongly support the use of systems for quarantine security for apples without need for a specific single treatment. Similar systems without the orchard control program are currently used for cherries shipped to Japan and Korea (Moffitt, 1990). Data gathered from the U.S. Department of Agriculture, Animal and Plant Health Inspection Service, Plant Protection and Quarantine (USDA-APHIS-PPQ), and the Pacific Northwest cherry industry on the incidence of codling moth larvae in shipments to Japan and Korea over an 11 year period showed that only two larvae in an estimated 316 million cherries have been found. Japan, however, continues to require a direct treatment such as methyl bromide as part of their certification of imported cherries. Both systems would meet or exceed the standard probit 9 (99.9968% mortality) level of security without the need for a specific postharvest treatment.

Shipment of Hawaii-Grown Papaya to the Mainland U.S.

Papaya grown in Hawaii can be infested with tephritid fruit flies such as Mediterranean fruit fly, *Ceratitis capitata* (Wiedemann), oriental fruit fly, *Bactrocera dorsalis* (Hendel), and melon fly, *Bactrocera cucurbitae* (Coquillett). For the past several years, papayas were allowed to enter the U.S. based on a system developed by Couey & Hayes (1986). Their system was composed of a preharvest fruit selection procedure based on an objective measurement of fruit color and a two stage hot water treatment to ensure that immature insects in the fruit would be killed. Then fruit was inspected visually prior to packing and certification. In developing this system, the authors referred to work by Couey et al. (1984) which reported that fruit less than one-quarter ripe was rarely infested in the field and showed that the two stage hot water treatment was sufficient to kill eggs near the surface of the fruit. The subsequent discovery of a blossom end defect in some fruits that allowed larvae to migrate deep inside the seed cavity (Zee et al. 1989) resulted in an additional visual inspection requirement to remove blossom end defective fruit prior to packing. This system used information on the infestation biology of the insects (Seo et al. 1982) as well as knowledge of the chemistry of the fruit (Seo et al. 1983) to minimize the chance of infested fruit arriving at the packing house. The two-stage hot water dip incorporated the standard dip for disease control with a second dip required for insect control.

Earlier, Couey et al. (1984) developed another quarantine treatment system based on the fruit selection criteria discussed above. With this system, the fruit selection procedure was followed by the standard hot water dip for disease control (49°C for 20 min) followed by an in transit cold treatment (8 - 9°C for 10 d). Final inspection and certification occurred at destination using information from temperature monitors within the shipping container.

Shipment of 'Sharwil' Avocados to the U.S. Mainland

A system has been used to meet quarantine requirements for the export of Hawaii-grown 'Sharwil' avocados to the mainland U.S. (Armstrong 1991). It consisted of harvesting only 'Sharwil' avocados with the stem attached, shipping to a certified fruit fly-free packing house within 12 h, sorting and culling procedures to remove damaged fruit prior to packing, final inspection, and certification. This approach was based on the fact that 'Sharwil' avocado is not a host of fruit flies as long as the fruit is attached to the tree (Armstrong et al. 1983, Armstrong 1991).

Several other examples of systems used to satisfy quarantine security have been reported. Yokoyama et al. (1992) recommended a system for the shipment of stone fruits to New Zealand which would ensure that fruits would not be infested with walnut husk fly. The system incorporates a fly free period during which early-season stone fruits could be harvested and exported prior to adult emergence. A methyl bromide fumigation would be used to ensure that no flies exist in fruit packed later in the season. Final inspection and certification would follow fumigation.

Curtis et al. (1991) reported that codling moth is rarely found in California-grown nectarines (based on preharvest trap surveys and postharvest monitoring of culled and packed fruits) and has calculated worst case estimates. The estimates show that the highest number of codling moth larvae that might be present in a given transoceanic van shipment (27,500 - 81,500 fruit per container) are 0.65 - 1.94 larvae per shipment. Based on these data, <24 codling moth would be found per one million nectarines 95% of the time, which would exceed the probit 9 level of quarantine security. They suggested a system for nectarines aimed at providing quarantine security without the need for a direct postharvest treatment. Their system included integrating inspection of harvested fruit on arrival at the packing house, with rejection of those lots not meeting established standards, grading, sorting, and packing procedures emphasizing removal of fruits damaged by codling moth, and final inspection/certification of nectarines in the box as packed for export.

Vail et al. (1993) proposed a system for ensuring quarantine security for codling moth inside in-shell walnuts. Their system incorporated information on infestation levels at harvest and thoroughly analyzed the postharvest biology and survival of the insect in relationship to the movement of the walnuts through normal marketing channels. Specific parameters studied for their impact on the probability of eventual survival of a mating pair of adults at a location included population of codling moth at harvest, age distribution and survival on harvested nuts, emergence patterns of male and female moths, mating behavior, and the myriad marketing factors that are involved in shipping walnuts. Each factor could impact the probability of insect survival for quarantine security.

Cowley et al. (1991) described a system that included a direct methyl bromide fumigation treatment to disinfest Tongan watermelons infested with *Bactrocera xanthodes* (Broun) for export to New Zealand. Their system incorporated biological information regarding the probable infestation level of export quality melons with a treatment sufficient to met New Zealand's quarantine security level (Baker et al. 1990). Baker (1990) also discussed the use of systems practices as part of New Zealand's bilateral quarantine agreements with countries importing fruit fly host material into that country. Systems practices are used to reduce the risk that a particular shipment of commodities arriving at a specific location on a given day does not exceed New Zealand's maximum pest limit (Baker et al. 1990).

Future Technology Using Systems

It is apparent from the above discussion that systems are complex and rely on sound information on the pest biology and how preharvest and postharvest operational parameters affect pest infestation levels to ensure that quarantine security can be achieved. Systems are being incorporated into the quarantine treatment programs as successful examples of systems and their use emerge. For systems to gain greater acceptance by regulatory decision makers, industry, and the public, progress must be made towards improving and quantifying knowledge of the complex interactions between the pests and their hosts. A realistic goal must be established to determine how these interactions integrate into the complex packing, shipping, and marketing channels that most commodities enter. This integration of information makes systems useful and unique.

Until recently, research focused on the uses of direct treatments having the ability to kill insects at any level of infestation. However, the narrow margin that separates the efficacy of many of these treatments against insects and the potential of damage to the commodity requires new analysis of conventional paradigms that rely solely on the use of direct quarantine treatments. Increased concern over the use of toxic compounds should serve to additionally focus attention on meaningful and effective alternatives. All available knowledge on the pest in question must be incorporated to develop viable alternatives to achieve quarantine security. What constitutes a quarantine risk should rely heavily on our knowledge of the biological interactions between the pest and its host rather than "perceived risk." The use of this knowledge will better define quarantine security through knowledge of the complex interactions that impact production, shipment, and marketing of commodities.

Acknowledgments

The authors thank A. L. Knight (USDA-ARS, Yakima, WA), J. W. Armstrong and N. J. Liquido (USDA-ARS, Hilo, HI), and R. Mangan (USDA-ARS, Weslaco, TX) for their helpful discussions. The authors also thank M. Purcell (USDA-ARS, Kapaa, HI), R. Messing (Univ. of Hawaii, Kapaa, HI), P. V. Vail (USDA-ARS, Fresno, CA), R. T. Baker (MAF, Auckland, New Zealand) and M. Kirby (APHIS, Guatemala City, Guatemala) for review of the manuscript.

References

Armstrong, J. W. 1991. 'Sharwil' avocado: Quarantine security against fruit fly infestation in Hawaii. J. Econ. Entomol. 84: 1308-1315.

Armstrong, J. W., W. C. Mitchell & G. J. Farias. 1983. Resistance of 'Sharwil' avocados at harvest maturity to infestation by three fruit fly species in Hawaii. J. Econ. Entomol. 76: 119-121.

Baker, A. C. 1939. The basis for treatment of products where fruit flies are involved as a condition for entry into the United States. USDA Circular 551.

Baker, R. T. 1990. The importation of fruit fly host material into New Zealand. Proceedings, Papua, New Guinea, First International Fruit Conference. Rabaul, 1989. Pp. 83-99.

Baker, R. T., J. M. Cowley, D. S. Harte & E. R. Frampton. 1990. Development of maximum pest limit for fruit flies (Diptera: Tephritidae) in produce imported into New Zealand. J. Econ. Entomol. 83: 13-17.

Couey, H. M. & C. F. Hayes. 1986. Quarantine procedure for Hawaiian papaya using fruit selection and a two-stage hot-water immersion. J. Econ. Entomol. 79: 1307-1314.

Couey, H. M., E. S. Linse & A. N. Nakamura. 1984. Quarantine procedure for Hawaiian papayas using heat and cold treatments. J. Econ. Entomol. 77: 984-988.

Cowley, J. M., R. T. Baker, K. G. Englberger & T. G. Lang. 1991 Methyl bromide fumigation of Tongan watermelons against *Bactrocera xanthodes* (Diptera: Tephritidae) and analysis of quarantine security. J. Econ. Entomol. 84: 1763-1767.

Curtis, C. E., J. D. Clark & J. S. Tebbets. 1991. Codling moth (Lepidoptera: Tortricidae) incidence in packed nectarine fruit. J. Econ. Entomol. 84: 1686-1690.

Jang, E. B. 1990. Fruit fly disinfestation of tropical fruits using semipermeable shrinkwrap films. Acta Horticulturae. 269: 453-458.

Knight, A. L. & H. R. Moffitt. 1991. Removal of codling moth injured apples in packinghouses. Proceedings, Washington State Hortic. Association (1990) 86: 211-213.

Moffitt, H. R. 1990. A systems approach to meeting quarantine requirements for insect pests of deciduous fruits. Proceedings, Washington State Hortic. Association (1989) 85: 223-225.

Seo, S. T. & C. S. Tang. 1982. Hawaiian fruit flies (Diptera: Tephritidae): Toxicity of benzyl isothiocyanate against eggs of first instars of three species. J. Econ. Entomol. 75: 1132-1135.

Seo, S. T., G. J. Farias & E. J. Harris. 1982. Oriental fruit fly: ripening of fruit and its effect on the index of infestation of Hawaiian papayas. J. Econ. Entomol. 75: 173-178.

Seo, S. T., C. S. Tang, S. Sanidad & T. H. Takenaka. 1983. Hawaiian fruit flies (Diptera: Tephritidae): variation of index of infestation with benzyl isothiocyanate concentration and color of maturing papayas. J. Econ. Entomol. 76: 535-538.

Vail, P. V., J. S. Tebbets & B. E. Mackey. 1993. Quarantine treatments: A biological approach to decision making for selected hosts of codling moth (Lepidoptera: Tortricidae). J. Econ. Entomol. 86: 70-75.

Yokoyama, V. Y., G. T. Miller & P. L. Hartsell. 1992. Pest-free period and methyl bromide fumigation for control of walnut husk fly (Diptera: Tephritidae) in stone fruits exported to New Zealand. J. Econ. Entomol. 85: 150-156.

Zee, F. T., M. S. Nishina, H. T. Chan & K. A. Nishijima. 1989. Blossom end defects and fruit fly infestation in papayas following hot water quarantine treatment. HortScience 24: 323-325.

16

Combination and Multiple Treatments

Robert L. Mangan and Jennifer L. Sharp

Combination/multiple treatments are serial applications of two or more single treatments, which alone do not achieve quarantine security (99.9968% mortality [Baker 1939]), but will when applied in combination. Treatments should be additive and possibly synergistic. Combination/ multiple treatments may be used effectively when a single treatment is not practical. Examples of a single treatment that are not practical include chemical treatment that produces unacceptable residues, vapor heat treatment that damages the condition or quality of the commodity at doses needed to reach quarantine security, and any expensive single treatment. The rationale for a combination/multiple treatment is based on the concept that pest survival in the commodity is reduced following each quarantine treatment. Combination/multiple treatments may be included as part of systems approaches, for example, as with a fly free certification program in Florida against Caribbean fruit fly, *Anastrepha suspensa* (Loew) (Chapter 14, Nguyen et al. 1992).

In this chapter we review the use of combination/multiple treatments to reduce the probability of pest infestation. Also, we discuss several statistical methods used to evaluate effectiveness of combination/multiple treatments.

Treatment Examples

Many examples of combination/multiple treatments that disinfest fruits of quarantine pests are found in the literature. The United States Department of Agriculture, Animal and Plant Health Inspection Service,

Plant Protection and Quarantine (USDA-APHIS-PPQ) Treatment Manual published many different treatment schedules for methyl bromide plus refrigeration that kill larvae of Mediterranean fruit fly, *Ceratitis capitata* (Wiedemann), and oriental fruit fly, *Bactrocera dorsalis* (Hendel), in citrus, stone fruits, pome fruits, and ethrog from different countries, and light brown apple moth complex, *Epiphyas* species (T108) (Anonymous 1992a). Methyl bromide followed by refrigeration also kills infesting larvae of Mediterranean fruit fly, oriental fruit fly, and melon fly, *Bactrocera cucurbitae* (Coquillett), in Hawaii-grown papaya, *Carica papaya* L., and avocado, *Persea americana* Mill., (Anonymous 1992a, Seo et al. 1971), and Caribbean fruit fly immatures in Florida-grown grapefruit (Benschoter 1982).

The USDA-APHIS-PPQ Treatment Manual published different treatment schedules for refrigeration followed by methyl bromide fumigation (T109). The treatments provided quarantine security against *Austrortrix* and *Epiphyas* species in apples and pears and larvae of Queensland fruit fly, *Bactrocera tryoni* (Froggatt), and Mediterranean fruit fly in grapes from Australia (Anonymous 1992b). Other examples of combination/multiple treatments are selection of fruit for degree of fruit ripeness and hot water immersion to disinfest Hawaii-grown papayas of fruit flies (Couey & Hayes 1986); oviposition preference, hot water immersion, and refrigeration to disinfest Hawaii-grown papayas of fruit flies (Couey et al. 1984); methyl bromide fumigation plus controlled atmosphere to disinfest apples of codling moth, *Cydia pomonella* (L.), (Moffitt 1971); irradiation plus refrigeration to disinfest Florida-grown grapefruit, *Citrus paradisi* Macf., of Caribbean fruit fly immatures (von Windeguth & Gould 1990); hot water immersion plus refrigeration to disinfest Caribbean fruit fly immatures in grapefruit (Gould 1988); and refrigeration plus controlled atmosphere to control a wide variety of pests in different commodities (Glass & Chapman 1961, Dustan 1963, Benschoter 1987, Toba & Moffitt 1991, Whiting et al. 1991). Yokoyama et al. (1993) developed a multiple quarantine treatment of compression and hydrogen phosphide fumigation (60 g per 28.3 cubic meters for 7 d at >10°C). The treatment provided quarantine security against introduction in Japan of Hessian fly, *Mayetiola destructor* (Say), in six species of hay. Commercial hay compressors produced 72 and 80 kg per square centimeter of pressure and caused 93 and 97% mortalities of Hessian fly puparia in bales held for 7 d in freight containers, respectively. Hydrogen phosphide fumigation significantly increased the mortality of puparia that survived compression in the bales. The effect of hydrogen phosphide alone (without bale compression) was not measured.

From the early 1970s to the 1980s, hot water dips followed by fumigation with ethylene dibromide (EDB) were used as combination treatments that reduced postharvest fruit rots and killed Tephritidae immatures in fruits. Seo et al. (1972) reported that eggs and larvae of Mediterranean fruit fly, oriental fruit fly, and melon fly were killed in mangoes sequentially immersed in 46.3°C water for 120 min, and then fumigated in wooden field boxes with EDB. A similar treatment was reported by Lin et al. (1976) who found no oriental fruit fly and melon fly larvae in Taiwan-grown mangoes immersed in 48 - 50°C water for 20 min, hydrocooled, dried and cooled, and then fumigated with EDB. Couey et al. (1985) reduced EDB levels necessary for quarantine security for papayas infested with oriental fruit fly, melon fly, and Mediterranean fruit fly from 8 - 4 g per cubic meter by fumigating the papayas after immersion in 49°C water for 20 min.

Statistical Approaches

Methods to analyze the effects of more than one treatment on pest mortality and interactions between treatments have been presented by Finney (1971), Couey & Chew (1986), and Robertson & Preisler (1992). The expected (calculated) mortality (MC2) from several treatments described in Finney (1971) and Robertson & Preisler (1992) using mortalities $t_1, t_2, t_3 \ldots$ is

$$MC2 = t_1 + (1 - t_1)t_2$$

for t_1 and t_2 combined. The expression can be arranged to give

$$MC2 = 1 - (1 - t_1)(1 - t_2)$$

For three combined treatments, the expression is

$$MC3 = 1 - (1 - t_1)(1 - t_2)(1 - t_3)$$

This expression can be expanded when additional treatments are included.

The expected mortality derived from the equations should be compared with the experimentally observed mortality to evaluate the additive effects of combined treatments. If more than two treatments are combined, observed and expected mortalities should be compared to evaluate the effectiveness of combined treatments. Comparisons of observed and calculated mortalities can be used to determine possible synergistic or antagonistic interactions among the treatments. Interactions may depend on changes in the physiology of the commodity or pest.

One of the first studies using combination/multiple treatments in which treatment effects were calculated separately was conducted by Seo et al. (1971). The study determined the separate effects of fumigation and refrigeration and the combined effects of three treatments including fumigation with methyl bromide and refrigeration before and after fumigation. Seo et al. (1971) presented computations for expected mortality from combined treatments. Equation no. 5 in that study is incorrect, and equation no. 6 was given as

$$f_1 = f_1 + (1 - f_1) + (1 - f_1 - f_2 + f_1 f_2) f_3$$

where f represents mortalities in decimal fractions for the combined, integrated treatment, and 1, 2, and 3 represent treatments 1, 2, and 3, respectively. Since

$$f_1 + (1 - f_1) = 1$$

the equation is identical to one presented in Finney (1971). The individual treatment mortalities in Seo et al. (1971) were estimated from regression equations rather than from direct measurement or estimation by comparing treated and control fruits.

Among the publications for combination/multiple treatments, four contained calculations or provided data to calculate the expected mortality of combined treatments from measured mortalities of individual treatments. von Windeguth & Gould (1990), Couey et al. (1984) and Seo et al. (1971) provided complete data sets. Moffitt (1971) provided mortality data for individual treatments and combinations of the treatments; however, one codling moth larvae survived treatment with 32 g per cubic meter of methyl bromide. The treatment dose was too low to permit meaningful calculations. Seo et al. (1972) provided mortality data for individual treatments and their combinations; however, the study combined the data for oriental fruit fly and Mediterranean fruit fly.

Three studies of combination/multiple treatments are examined here for synergistic effects. Data are presented in Table 16.1 (von Windeguth & Gould 1990), Table 16.2 (Seo et al. 1971), and Table 16.3 (Couey et al. 1984). The estimates of synergism among the treatments were made by comparing the estimate of the true mortality and the calculated, expected mortality based on mortality estimates from each treatment. Percentage synergism was erratic for combined treatments when percentage mortality for one or both treatments alone was <90%. The result may have been due to large errors in estimating percentage mortalities, since

TABLE 16.1 Calculated and estimated mortalities for gamma radiation and cold storage as single and combination quarantine treatments against *Anastrepha suspensa* in grapefruit (von Windeguth & Gould 1990)

Treatment 1	Treatment 2	Larval Survival	Estimated % Mortality[a]	Calculated % Mortality[b]	% Synergism
Cold storage (Days 1.1°C)	Irradiation (Grays)				
0	0	7,567	0	--	--
2	0	1,472	60.92	--	--
4	0	261	93.07	--	--
6	0	51	98.65	--	--
0	0	3,808	0	--	--
0	5	3,692	3.04	--	--
0	10	3,672	3.57	--	--
0	20	873	77.07	--	--
0	40	445	88.31	--	--
0	80	252	98.38	--	--
0	0	3,097	0	--	--
2	5	1,596	48.46	62.11	-28.16
2	10	859	72.26	62.23	+16.12
2	20	423	86.34	91.03	-05.16
2	40	534	82.76	93.26	-11.25
2	80	374	87.92	99.36	-11.51
0	0	1,153	0	--	--
4	5	114	90.11	93.28	-03.39
4	10	266	76.93	93.31	-17.56
4	20	23	98.77	98.41	+00.36
4	40	30	97.40	99.19	-01.80
4	80	3	97.93	99.84	-02.12
0	0	2,828	0	--	--
6	5	245	99.15	98.65	+00.51
6	10	12	99.58	98.65	+00.94
6	20	7	99.78	99.69	+00.09
6	40	3	99.89	99.84	+00.05
6	80	0	100.00	99.98	+00.02

[a] Mortality determined as numbers of 1 - survivors in treated/control fruit.

[b] Expected mortality calculated from separate effects of gamma radiation and cold storage.

TABLE 16.2 Calculated (from separate treatments) and estimated mortalities for methyl bromide and cold storage as quarantine treatments for *Ceratitis capitata* in Hawaii-grown papayas (Seo et al. 1971)

Treatment 1	Treatment 2	Estimated % Mortality[a]	Calculated % Mortality[b]	% Synergism
Fumigation Time (32g/m³ MeBr)	Cold Storage 4.4°C (days)			
2.0	0	99.68	--	--
2.5	0	99.09	--	--
0	3	55.01	--	--
0	4	69.14	--	--
0	6	84.55	--	--
0	9	93.81	--	--
0	11	93.38	--	--
2.0	3	99.87	99.71	−00.16
2.0	4	99.90	99.98	+00.08
2.0	9	99.98	99.99	+00.01
2.0	11	99.99	100.00	+00.01
2.5	3	99.59	99.99	+00.40
2.5	6	99.86	100.00	+00.14

[a] Mortality determined as numbers of survivors in treated control fruit.
[b] Expected mortality calculated from separate effects of methyl bromide and cold storage.

TABLE 16.3 Calculated and observed percentage mortalities for hot water dip and cold storage as single and combination quarantine treatments for *Bactrocera dorsalis* disinfestation from Hawaii-grown papayas (Couey et al. 1984)

Treatment	Time	Pupae surviving	Estimated Population	Estimated % Mortality[a]	Calculated % Mortality[b]	% Synergism
Control		18,865				
Heat 49°C	20 Min	615	18,602	96.693		
Cold 10°C	5 d	277	18,490	99.501		
Heat + Cold		75	18,208	99.588	99.950	−0.362
Control		27,786				
Heat 49°C	20 Min	1,150	28,659	95.987		
Cold 10°C	7 d	75	28,130	99.733		
Heat + Cold		1	27,760	99.996	99.989	+0.007

[a] Mortality determined as numbers of survivors in treated/control fruit.
[b] Expected mortality calculated from separate effects of hot water dip and cold storage.

the infestation rate of the treated fruit had to be calculated from survival in control fruit. The estimate of synergism was positive in studies by von Windeguth & Gould (1990) and Seo et al. (1971) when individual mortalities were >90%. The variation in the percentage synergism cannot be evaluated because no variance was given in either infestation among fruit in the control or survivorship among the treated fruit. Data showed no detectable antagonism between the treatments as mortalities reached 99.9968%.

In contrast to the studies that allowed calculation of expected mortalities from individual treatment data, other studies presented data for various groups of combination/multiple treatments and did not include data for individual treatments. Often, the tests were done using ranges of treatment intensities from previous experiments. For example, Benschoter (1982) used different doses of methyl bromide in combination with three weeks of refrigeration. Benschoter (1979) used ranges of methyl bromide doses that had been tested for optimum mortality effects. The degree of mortality met quarantine security; however, the separate mortality data for the two doses were not presented. Thus, interactions of methyl bromide and refrigeration treatments could not be calculated.

Conclusions

The use of combination/multiple treatments to disinfest commodities of pests is an effective quarantine treatment application when one treatment alone is unsatisfactory. Major considerations are effects on the quality of the treated commodity and effectiveness of the treatments to kill pests.

Two general approaches to the development of combination/ multiple treatments are discussed in this review. One approach involved testing the treatments individually, calculating an expected mortality from the individual mortality data, then confirming the predicted mortality at the 99.9968% mortality level. Another approach predicted optimum combinations of treatments from published research reports in the review. Various treatment combinations were tested for optimum combined effect and then confirmed for the best combination. The recommended combination usually was based on biological efficacy (acceptable pest mortality and commodity quality) and the cost of the treatment.

The tabulated data for the expected and observed mortality rates showed that the expected and observed values at 50 - 90% mortalities varied considerably. Analysis of the data did not indicate strong synergism or antagonism among treatments.

Synergistic interactions among treatments could reduce costs or reduce damage to the fruit quality. Tropical fruits that could be damaged by single treatments such as gamma irradiation, vapor heat, or hot water immersion perhaps could be treated with combination/multiple treatments which would provide quarantine security and not damage the treated commodity.

References

Anonymous. 1992a. Animal and Plant Health Inspection Service. Plant protection and quarantine treatment manual. T 108 - Fumigation plus refrigeration of fruits. U.S. Government Printing Office. Washington, D.C.

_____. 1992b. Animal and Plant Health Inspection Service. Plant protection and quarantine treatment manual. T109 - Refrigeration plus fumigation of fruits. U.S. Government Printing Office. Washington, D.C.

Baker, A. C. 1939. The basis for treatment of products where fruitflies are involved as a condition for entry into the United States. USDA Circular 551.

Benschoter, C. A. 1979. Fumigation of grapefruit with methyl bromide for control of *Anastrepha suspensa*. J. Econ. Entomol. 72: 401-402.

_____. 1982. Methyl bromide fumigation followed by cold storage as a treatment for *Anastrepha suspensa* (Diptera: Tephritidae) in grapefruit. J. Econ. Entomol. 75: 860-862.

_____. 1987. Effects of modified atmospheres and refrigeration temperatures on survival of eggs and larvae of the Caribbean fruit fly (Diptera: Tephritidae) in laboratory diet. J. Econ. Entomol. 80: 1223-1225.

Couey, H. M. & V. Chew. 1986. Confidence limits and sample size in quarantine research. J. Econ. Entomol. 79: 887-890.

Couey, H. M. & C. F. Hayes. 1986. Quarantine treatment for Hawaiian papaya using fruit selection and a two-stage hot-water immersion. J. Econ. Entomol. 79: 1307-1314.

Couey, H. M., E. S. Linse & A. N. Nakamura. 1984. Quarantine procedure for Hawaiian papayas using heat and cold treatments. J. Econ. Entomol. 77: 984-988.

Couey, H. M., J. W. Armstrong, J. W. Hylin, W. Thornburg, A. N. Nakamura, E. S. Linse, J. Ogata & R. Vetro. 1985. Quarantine procedure for Hawaiian papaya using a hot-water treatment and high-temperature, low dose ethylene dibromide fumigation. J. Econ. Entomol. 78: 879-884.

Dustan, G. G. 1963. The effect of standard cold storage and controlled atmosphere storage on survival of larvae of the oriental fruit moth, *Grapholitha molesta*. J. Econ. Entomol. 56: 167-169.

Finney, D. J. 1971. *Probit Analysis, 3rd Ed.* Cambridge University Press.

Glass, E. H., P. J. Chapman & R. M. Smoek. 1961. Fate of apple maggot and plum curculio larvae in apple fruits held in controlled atmosphere storage. J. Econ. Entomol. 54: 915-918.

Gould, W. P. 1988. A hot water/cold storage quarantine treatment for grapefruit infested with the Caribbean fruit fly. Proceedings, Fla. State Hortic. Soc. 101: 190-192.

Lin, T. H., F. C. Tseng, C. R. Chang & L. Y. Wang. 1976. Multiple treatment for disinfesting oriental fruit fly in mangoes. Plant Protection Bul. 18: 231-241.

Moffitt, H. R. 1971. Methyl bromide fumigation combined with storage for control of codling moth in apples. J. Econ. Entomol. 64: 1258-1260.

Nguyen, R., C. Poucher & J. R. Brazzel. 1992. Seasonal occurrence of *Anastrepha suspensa* (Diptera: Tephritidae) in Indian River County, Florida, 1984-1987. J. Econ. Entomol. 85: 813-820.

Robertson, J. L. & H. K. Preisler. 1992. *Pesticide Bioassays with Arthropods.* Boca Raton, Florida: CRC Press.

Seo, S. T., R. M. Kobayashi, D. L. Chambers, L. F. Steiner, J. W. Balock, M. Komura & C.Y.L. Lee. 1971. Fumigation with methyl bromide plus refrigeration to control infestations of fruit flies in agricultural commodities. J. Econ. Entomol. 64: 1270-1274.

Seo, S. T., D. L. Chambers, E. K. Akamine, M. Komura & C.Y.L. Lee. 1972. Hot water-ethylene dibromide fumigation-refrigeration treatment for mangoes infested with oriental and Mediterranean fruit flies. J. Econ. Entomol. 65: 1372-1374.

Toba, H. H. & H. R. Moffitt. 1991. Controlled-atmosphere cold storage as a quarantine treatment for nondiapausing codling moth (Lepidoptera: Tortricidae) larvae in apples. J. Econ. Entomol. 84: 1316-1319.

von Windeguth, D. L. & W. P. Gould. 1990. Gamma irradiation followed by cold storage as a quarantine treatment for Florida grapefruit infested with Caribbean fruit fly. Fla. Entomologist 73: 242-247.

Whiting, D. C., S. P. Foster & J. H. Maindonald. 1991. Effects of oxygen, carbon dioxide, and temperature on the mortality responses of *Epiphyas postvittana* (Lepidoptera: Tortricidae). J. Econ. Entomol. 84: 1544-1549.

Yokoyama, V. Y., J. H. Hatchett & G. T. Miller. 1993. Hessian fly (Diptera: Cecidomyiidae) control by hydrogen phosphide fumigation and compression of hay for export to Japan. J. Econ. Entomol. 86: 76-85.

17

Quality and
Condition Maintenance

Roy E. McDonald and William R. Miller

Horticultural commodities can be damaged during any handling or treatment subsequent to harvest. This is particularly true when commodities are subjected to some quarantine treatments required for disinfestation purposes. Quarantine treatments must be efficacious and must not adversely affect the commodity's quality, condition, and susceptibility to decay. If the quarantine treatment reduces the value of the commodity, then the treatment is not fully effective. Any treatment that disinfests a commodity should have minimal deleterious effects on that commodity. Damage manifests itself as the loss of market quality attributes including shelf life, appearance, flavor, texture, aroma, and increased susceptibility to decay organisms. Reduction in market quality impacts consumer acceptability. Return sales may be reduced and consumers may not be willing to pay as high a price for lower quality. Competition from sources not requiring quarantine procedures may take an advantaged position.

A number of postharvest treatments have been studied for the purpose of insect disinfestation of fresh fruits and vegetables. The responses of fruits and vegetables to quarantine treatments with respect to damage are presented in Table 17.1 at the end of this chapter. For purposes of this chapter, we review only studies that considered the resulting market quality of the product as a consequence of a disinfestation treatment. Ethylene dibromide fumigation is omitted because it is not approved for use on fruits and vegetables for human consumption (Ruckelshaus 1984).

Film Packaging

Gould & Sharp (1990a) found that mangoes deteriorated within 6 d of film packaging in two layers of Clysar EHC-150 film to disinfest fruit of Caribbean fruit fly, *Anastrepha suspensa* (Loew). Each mango was wrapped twice and the mangoes shriveled as they decomposed. Sharp (1990b) reported that individually wrapped grapefruit in Clysar EHC-50-F or double sealed in Clysar EHC-150-F film remained intact and did not collapse as they became senescent. Film packaging alone does not look promising as a quarantine treatment.

Fumigation

Fumigation with methyl bromide has been used on several cultivars of apples shipped from Japanese beetle, *Popillia japonica* Newman, quarantine areas in the United States in the 1930s and 1940s. Observations by Kenworthy (1944) of 44 cars of apples shipped from Delaware revealed that methyl bromide caused injury to some cultivars of apples and not to others. 'Williams' apples were severely injured, and a higher fruit temperature at the time of fumigation caused greater surface injury (scald). Initial internal breakdown, manifested as browning of the carpel walls, was present in a larger percentage of the fruits when the fruit temperature was higher. Advanced internal injury, evidenced as browning of the flesh, tended to increase rapidly after removal from the iced cars. Subsequent studies by Kenworthy & Gaddis (1946) determined the tolerance of several cultivars of Delaware-produced apples to methyl bromide. They found six of 18 cultivars sustained damage and that symptoms varied with cultivar. No injury occurred on 'Cortland,' 'Jonathan,' 'Starking,' 'Blaxtayman,' 'Rome,' 'Gallia Beauty,' 'Lily of Kent,' 'Stark,' 'Paragon,' 'Gano,' 'Winesap,' or 'Neno.'

Claypool & Vines (1956) extensively tested fumigants on several cultivars of deciduous fruits and reported that four chemicals impaired appearance, flavor, or keeping quality only slightly. They found methyl bromide and methyl iodide negatively influenced the flavor and condition (dependent on cultivar and dose) of apple, apricot, cherry, fig, grape, nectarine, peach, pear, persimmon, plum, raspberry, and strawberry. Ethylene chlorobromide caused off-flavor and appearance problems in some cultivars of apple, grape, peach, plum, and strawberry (Claypool & Vines 1956).

Lindgren & Sinclair (1951) reported the tolerance of citrus and avocados to fumigants effective against oriental fruit fly, *Bactrocera dorsalis*

(Hendel). In general, navel oranges were more susceptible to injury than oranges, grapefruit, and lemons. No citrus cultivar was injured by fumigation with ethylene chlorobromide or methyl bromide at 16 g per cubic meter for 2 h at 27°C. Methyl bromide at 32 g per cubic meter bordered between injurious and noninjurious, depending upon the condition of the fruit. 'Fuerte' and 'Dickinson' avocados were severely injured by fumigation with ethylene oxide, acrylonitrile, and a mixture of 50% acrylonitrile and 50% carbon tetrachloride at 16 - 80 g per cubic meter for 2 h at 27°C (Lindgren & Sinclair 1951). These fruits were also susceptible to injury by methyl bromide at 32 g per cubic meter for 4 h.

Drake et al. (1988) considered fruit quality as a result of methyl bromide fumigation of three cultivars of apples and the impact of apple waxes to retain quality. Apples were fumigated with methyl bromide at 16, 32, or 48 g per cubic meter at 6 and 20°C and at 56 g per cubic meter at 6°C for 2 h. They reported that methyl bromide fumigation for control of codling moth, Cydia pomonella (L.), did not detract from firmness and skin color, but could result in fruit with low quality, particularly with respect to internal color. Also, compared with unwaxed apples, differences in the reaction of waxed apples to methyl bromide appeared to be related to the absorption or desorption of methyl bromide to or from the wax material.

Anthon et al. (1975) controlled codling moth in sweet cherries with methyl bromide at 32 g per cubic meter for 2 h at 24°C with no significant effects on quality or taste of the treated fruits. Methyl bromide fumigation at 32 g per cubic meter for 4 h or at 48 g per cubic meter for 2 h at 19°C to disinfest cucumbers of melon fly, Bactrocera cucurbitae (Coquillett), and oriental fruit fly did not produce phytotoxic effects sufficient to affect marketability (Armstrong & Garcia 1985). No visible damage was observed, and there was no loss of crispness or flavor. Harvey et al. (1989) found that methyl bromide at 48 g per cubic meter for 2 h at 21°C to control codling moth in six cultivars of nectarines caused no significant phytotoxic response in any of the cultivars. No detrimental effects of the methyl bromide quarantine treatment at the same dose were found on 'Royal Giant' nectarines (Yokoyama et al. 1990).

Grapefruit fumigated with phosphine to control Caribbean fruit fly had significantly more rind breakdown following four weeks of storage compared with control fruit (Hatton et al. 1982a). The rind injury was manifested as aging and pitting. Rind scald occurred only on fruit that was fumigated under refrigerated conditions. They reported significantly more decay in fruit fumigated under refrigerated conditions compared with nonrefrigerated fruit.

Hallman & King (1992) controlled Caribbean fruit flies in carambolas with methyl bromide at 40 g per cubic meter for 2 h at 23°C. However, the shelf life at room temperature was reduced by 24 - 30% due to an increased rate of ripening and accelerated decomposition.

Hot Water Immersion

Armstrong (1982) showed that unripe 'Brazilian' bananas tolerated a 15 min, 50°C hot-water immersion treatment without loss of quality, whereas the riper stages sustained some peel scald as a treatment against Mediterranean fruit fly, *Ceratitis capitata* (Wiedemann), melon fly, and oriental fruit fly.

A two-stage hot-water immersion treatment was developed by Couey & Hayes (1986) against melon fly, Mediterranean fruit fly, and oriental fruit fly for Hawaii-grown papaya. Fruit less than one-quarter ripe was immersed in 42°C water for 30 or 40 min, followed immediately by a second immersion for 20 min in 49°C water. Papayas ripened normally if treatments were carefully applied. This treatment was discontinued because a blossom-end deformity that occurred in some Hawaii-grown papayas beginning about 1987 allowed the survival of fruit fly larvae in fruit that met the color standards. The two-stage hot-water immersion treatment was rescinded 1 July 1992 (Anonymous 1991).

Immersing both tart ('Golden Star' and 'Star King') and sweet ('Arkin,' 'Demak,' 'Butts Dwarf,' 'Key West,' 'Sri Kembayan,' 'Fwang Tung,' 'B-10,' 'Hart,' and 'Wheeler') carambolas in 46 - 46.4°C water for 55 - 85 min adversely affected market quality (Hallman 1989, Hallman & Sharp 1990). A few days after treatment, tart carambolas turned a dull yellow and became soft. Shelf life of sweet carambolas was reduced 2 - 4 d by the treatment. Hallman (1991) reported that water immersion at 49 - 49.3°C for 25 or 35 min resulted in excessive damage to carambolas, but that there was no difference in marketability of fruit treated at 43.3 - 43.6°C for 55 or 70 min or at 46 - 46.3°C for 35 or 45 min compared with controls.

Sharp (1990a) found that immersing California nectarines, plums, and peaches in 49.4 - 50°C water for 5 - 25 min resulted in unacceptable market quality. Regardless of cultivar, the treated stone fruits displayed unacceptable amounts of pitting, shriveling, surface scald, and bland flavor compared with control fruit.

Sharp (1985) reported that 'Marsh' grapefruit immersed in 48.9°C water for 20 min or longer exhibited severe scalding and pitting and produced off-flavors compared with control fruit. Miller et al. (1988, 1989) found that immersing 'Marsh' grapefruit in 43.5°C water for 4 h and 30 min caused peel discoloration, puffiness, and decreased resistance of peel to penicillium infection. McGuire (1991b) also reported increased decay caused by penicillium infection when 'Marsh' grapefruit was immersed in water at either a constant 48°C for 2 h or with a gradual increase to 48°C lasting 3 h. The immersion at a constant 48°C significantly increased weight loss and promoted peel injury while reducing firmness and color intensity.

Immersion of 'Tommy Atkins' and 'Keitt' mangoes in 46.1°C water for 65 min reduced the incidence of mango stem-end rot and anthracnose without significantly affecting fruit quality (Sharp & Spalding 1984). McGuire (1991a) also reported a significant decrease in stem-end rot and anthracnose without reducing market quality after immersion of 'Tommy Atkins,' 'Keitt,' and 'Palmer' mangoes in 46°C water for 90 - 115 min. An immersion for 150 min in which the water temperature gradually rose to 48°C, however, had no effect on decay reduction and did not reduce quality (McGuire 1991a).

Lenticels were darker on 'Tommy Atkins' immersed in 46°C water for 120 min, on 'Keitt' immersed in 46°C water for 90 min, and on both cultivars immersed in 49°C water for 60 min (Spalding et al. 1988). However, immersing mangoes in 46°C water for 60 - 90 min did not reduce market quality (Spalding et al. 1988). In subsequent tests, Sharp (1986) reported immersion in 46.1 - 46.7°C water for 45 - 65 min produced no visible injury to 'Tommy Atkins' or 'Keitt' mangoes. Segarra-Carmona et al. (1990) also reported no damage to 'Keitt' mangoes immersed in 46.1 - 46.7°C water for 60 min. They reported that mangoes picked too green shriveled and displayed pitted areas on the skin and the treatments appeared to increase visibility of fruit blemishes. Immersion of 'Tommy Atkins' and 'Keitt' in 46.1 - 46.7°C water for 90 min did not reduce fruit quality (Sharp et al. 1989b).

Sharp et al. (1989c) reported that immersion of Mexico-grown 'Ataulfo' mangoes in 45.9 - 47.1°C water for 90 min for quarantine security against Mediterranean fruit fly and *Anastrepha serpentina* (Wiedemann) did not reduce fruit quality. Immersion of Mexico-grown 'Oro,' 'Haden,' 'Keitt,' 'Tommy Atkins,' and 'Kent' mangoes in 46.1°C water for 90 min for quarantine security against Mexican fruit fly, *Anastrepha ludens* (Loew), and West Indian fruit fly, *Anastrepha obliqua* (Macquart), did not adversely affect fruit quality (Sharp et al. 1989a). However, the percentage of acceptable 'Francis' mangoes produced in Haiti decreased as the length of immersion time at 46.1 - 46.7°C increased beyond 75 min (Sharp et al. 1988).

Jacobi & Wong (1991) used water immersions ranging from 48 - 50°C for 20 min to disinfest 'Kensington' mangoes of Queensland fruit fly, *Bactrocera tryoni* (Froggatt). The fruits were injured when exposed to temperatures between 52 and 56°C, including scald, damaged lenticels, and disease development. However, treatment at 52°C enhanced color development when the fruit ripened. Treatment between 48 and 52°C increased the preclimacteric rise in respiration and shortened the time to fruit softening.

Hot water immersions at temperatures needed to disinfest cucumbers of melon fly, oriental fruit fly, and Mediterranean fruit fly caused surface pitting and yellowing of the fruit. Chan & Linse (1989) showed that conditioning 'Burpee Hybrid II' cucumber at 32.5°C for 24 h increased their tolerance to immersions in 45°C water for 30 min to at least 60 min, and at 46°C from <30 min to at least 50 min.

Immersing 'Ruby' guavas in 46.1°C water for 31 min provided probit 9 quarantine security against Caribbean fruit fly and did not reduce fruit quality (Gould & Sharp 1992).

Vapor Heat

Baker (1952) reported that the market quality of citrus treated with vapor heat at 43.3°C for 8 h was not reduced while disinfesting the fruit of Mediterranean fruit fly eggs and larvae.

Other early work with vapor heat for insect disinfestation was performed by Sein (1935) in Puerto Rico who reported that 8 h at 43°C in a circulating atmosphere saturated with moisture did not alter the flavor, texture, or keeping quality of 'White' mangoes probably infested with West Indian fruit fly and guavas infested with Caribbean fruit fly if the mangoes were held in refrigerated storage. Vapor heat at 43.5 ± 0.5°C for 3 h applied to Japan-grown mangoes for melon fly disinfestation did not injure the mangoes, but treatment at 47.5 ± 0.5°C for 3 h caused fruit damage (Sunagawa et al. 1987).

To disinfest 'Solo' papayas of Mediterranean fruit fly, Jones et al. (1939) concluded that a conditioning period of 6 h at 37.8°C and 60% relative humidity (RH) followed by about 2 h and 30 min at 43.3°C and 60% RH before a vapor heat treatment resulted in minimal fruit injury. Jones et al. (1939) treated papaya for 8 h at 43.3°C and 100% RH and reported that the flavor, aroma, color, and texture of treated fruit were not impaired. In general, ripe fruits were less tolerant than mature green fruit. Jones (1940) used the same vapor heat treatment for several Hawaii-produced vegetables, and preliminary data indicated that most treated vegetables sustained some damage.

Sinclair & Lindgren (1955) conducted one of the first comprehensive studies of the effect of vapor heat on the resulting condition and quality of fruit. Two vapor heat treatments against oriental fruit fly were employed with California citrus and avocado fruits: 1) an approach period in a saturated atmosphere of 5 h until the product reached 43.3°C, and 2) the "quick run up" method exposing the commodity to saturated vapor at 48.9°C until the product reached an internal temperature of 43.3°C. 'Fuerte' and 'Dickenson' avocados, lemons, and navel oranges were damaged by the vapor heat treatments. 'Valencia' oranges were not significantly damaged by either vapor heat treatment. Grapefruit was not seriously injured by the vapor heat treatment at 43.3°C for 8 h, but was significantly injured by the "quick run up" treatment at 48.9°C. However, the vapor heat treatments altered the taste of the citrus fruits and produced off-flavors which could be detected by a taste panel.

Hallman (1990) reported that carambolas treated at 49 - 49.3°C for 45 or 60 min showed extensive scalding after 2 - 3 d and were unmarketable. Vapor heat at 49 - 49.3°C for 45 or 60 min caused considerable brown spotting and scalding of carambolas after treatment and the fruits were considered unmarketable (Hallman 1991). Carambolas treated with vapor heat at 43.3 - 43.6°C for 90 or 120 min and at 46 - 46.3°C for 60 or 90 min appeared slightly darker and duller, and rib margins were darker compared with control fruit. Treated fruits were marketable (Hallman 1990).

Hallman et al. (1990) reported that vapor heat at 46 - 46.4°C for 3 h and 45 min resulted in darkening of the oil glands in the peel of 'Marsh' grapefruit and increased the susceptibility of the fruit to storage decay. However, in a nonreplicated test, vapor heat at 43.3 - 43.7°C for 5 h caused no damage to freshly harvested grapefruit (Hallman et al. 1990). Miller et al. (1989) reported development of scald on vapor heat-treated fruit at 43.5°C for 4 h and 25 min in a preliminary, nonreplicated test. However, in later replicated tests, Miller et al. (1991b) reported that similar vapor heat conditions actually reduced peel pitting five-fold compared with control fruit after five weeks of storage and did not cause peel discoloration or rind breakdown. 'Marsh' and 'Ruby Red' grapefruits did not develop symptoms of quality deterioration following a vapor heat treatment of 43.5°C for 4 h and 20 min for quarantine security against Caribbean fruit fly (Miller & McDonald 1991). An added benefit of the vapor heat treatment was that decay was reduced from about 5 to 2%, although aging was increased in the 'Marsh' cultivar. Miller & McDonald (1992) found that a vapor heat treatment (43.5°C for 4 h) was not detrimental to the quality of early-season 'Marsh' and 'Ruby Red' grapefruits that had been degreened in an ethylene chamber for 72 h before treatment. Symptoms of aging averaged 7% on nontreated fruit, and the vapor heat treatment was found to reduce it by 45% following five weeks of storage.

Sugimoto et al. (1983) found an approach period of about 2 h at 43.9 ± 0.3°C at 100% RH followed by a vapor heat treatment for 3 h at 43.4 ±

0.3°C for disinfestation of green pepper of oriental fruit fly caused no fruit damage. However, longer treatment times damaged the fruit.

Furusawa et al. (1984) reported no apparent damage on eggplant following a vapor heat treatment of 43.5 ± 0.5°C preceded by an approach time of 70 min to raise the fruit temperature to 43°C for disinfesting fruit of melon fly.

Forced Heated Air

Armstrong et al. (1989) reported that a four-stage forced heated air treatment consisting of 43 ± 1, 45 ± 1, 46.5 ± 1, and 49 ± 0.5°C air over 'Solo' papaya surfaces until the fruit core temperatures at the end of each temperature stage reached 41 ± 1.5, 44 ± 1, 46.5 ± 0.75, and 47.2°C, respectively, would disinfest fruit of Mediterranean fruit fly, melon fly, and oriental fruit fly. When fruit core temperatures reached 47.2°C, the papayas were immediately hydrocooled until core temperatures were <30°C. The fruit ripened normally with no reduction in fruit quality. Perhaps the fruit ripened normally because hydrocooling allowed the fruit to recover from any physiological disruption of the ethylene forming enzyme system caused by the heat treatment (Chan 1988, Armstrong et al. 1989). Chan et al. (1981) previously reported that extended treatments using hot water caused delayed softening of the fruit tissue which was correlated with a decrease in polygalacturonase activity.

Sharp (1989) found that the quality of 'Marsh' grapefruit treated with forced heated air at 46°C for 3 h was not reduced, and no differences were observed between treated and control fruits. McGuire (1991b) reported that no loss in quality resulted from several heat treatments of 'Marsh' grapefruit at 48°C for 3 h.

To establish the threshold boundary for stress tolerance of carambolas to forced heated air, Miller et al. (1990) treated fruit at 47, 48, and 49°C for 90, 120, or 150 min and found that treated carambolas deteriorated more rapidly, lost more weight, had more stem-end breakdown and rib browning, and generally had a more undesirable flavor than nontreated fruit. Treatment temperatures >47°C were unacceptable at the treatment durations of 90, 120, or 150 min. Treatment at 47°C for 90 - 120 min may be near the threshold for stress tolerated by carambolas.

McGuire (1991a) found that anthracnose was reduced and no injury developed on 'Keitt' and 'Palmer' mangoes treated with forced heated air at 46°C for 195 min or 48°C for 150 min. Miller et al. (1991a) reported that mangoes treated with forced heated air at 51.5°C for 125 min lost 1% more fresh weight than nontreated fruit and developed trace amounts of peel pitting. Total soluble solids concentrations and peel color at the soft-ripe stage were similar for treated and nontreated fruits (Miller et al. 1991a). Treated fruit generally reached the soft-ripe stage about a day earlier than nontreated fruit and had a lower incidence and severity of stem-end rot and anthracnose.

E. Mitcham & R. McDonald (unpublished data) found that similar treatment of 'Keitt' and 'Tommy Atkins' mangoes with forced heated air at 51.5°C for 125 min reduced the rate of respiration of the fruit; however, the respiratory climacteric occurred at the same time for control fruit. Heated fruit also softened at a reduced rate compared with controls and had a much lower incidence of anthracnose decay. These effects may be beneficial to the postharvest shelf life of mango.

Cold

Hatton & Cubbedge (1982) established that 'Marsh' and 'Ruby Red' grapefruits could be stored at 1°C for 21 d without suffering chilling injury if the fruits were first temperature conditioned for 7 d at 10, 16, or 21°C. Further research showed that conditioning grapefruit at 21 or 27°C for 7 d was significantly less effective than conditioning for a similar period of time at 16°C (Hatton & Cubbedge 1983). Working with Israel-grown 'Marsh' grapefruit, Chalutz et al. (1985) found that conditioning at 17°C for 6 d reduced chilling by more than 50% when the fruits were subsequently subjected to the cold treatment at 0°C for 10 d or at 2.2°C for 16 d. The presence of the fungicide thiabendazole in the wax coating of the fruit reduced the incidence of chilling injury by the same extent. The effect of combining a thiabendazole treatment with temperature conditioning was additive, and the susceptibility of grapefruit to chilling injury was reduced. Therefore, the cold treatment as a quarantine treatment can be practiced with a low risk of chilling injury and subsequent decay development.

Houck et al. (1990) found that conditioning 'Lisbon' lemons at 15°C for 7 d significantly reduced the incidence and severity of chilling injury when the fruits were held at 1°C for 14 or 28 d for tephritid fruit fly quarantine purposes. Rind injury that developed on the conditioned and cold-treated fruit was within acceptable levels for mid- and late-season fruits and may also be acceptable for the early-season lemons if the slight injury to the peel is disregarded.

Gould & Sharp (1990b) developed a cold-storage quarantine treatment for 'Arkin' carambolas where probit 9 was estimated to occur in 13.6 d at 1.1°C and in 21.1 d at 5°C and found that storage at 1.1°C for 15 d caused little or no damage to the fruit. However, in commercial practice damage has occurred (Craig Campbell, personal communication). In an attempt to reduce damage resulting from the cold treatment, Miller et al. (1991c) subjected 'Arkin' fruit to a conditioning treatment and surface coatings. They found that film wrapping carambolas significantly improved the maintenance of most quality attributes. Fruit appearance was improved due to a reduction in bronzing, weight loss was reduced nine-fold, and fruit remained very firm with a corresponding near-elimination of stem-end breakdown and shriveling.

Comparison of Temperature Management Treatments

The success of vapor heat, hot water immersion, or forced heated air seems to vary with commodity due to physical, physiological and/or morphological differences among commodity types. For example, peel damage of grapefruit is less severe with vapor heat (Miller & McDonald 1991) and forced heated air (McGuire 1991b) than with hot water immersion (McGuire 1991b). Core target temperatures for grapefruit are reached more rapidly with vapor heat than with forced heated air. Less damage is probably a result of the difference in the higher heat transfer coefficient of moist air versus that of air containing less moisture.

However, with papaya, treatment with vapor heat was found to result in nonuniform ripening, whereas the forced heated air treatment allowed normal ripening. 'Tommy Atkins' mangoes have shown fewer symptoms of stress following treatment with forced heated air compared with vapor heat. However, the successful forced heated air treatment of 'Tommy Atkins' mangoes caused serious damage to both the peel and pulp of 'Keitt' mangoes. Treatment response may vary with cultivar within a commodity.

Peel injury symptoms as a result of damage by heat or cold quarantine treatments usually differ. Cold sensitive tissue may develop the symptoms of pitting and discoloration of the peel. Peel damage due to heat may result in atypical color, but usually no pitting develops. Fruit subjected to quarantine treatments that employ heat should be free of bruising and other surface injuries that may occur during pretreatment handling. Heat treatments will enhance browning of injured tissue. Although the symptoms of injury may differ in severity or in expression by type of treatment, injury usually renders the tissue more susceptible to invasion by disease organisms or rejection by consumers.

Conditioning a commodity before exposure to a subsequent cold or heat quarantine treatment has been beneficial in the reduction or elimination of peel damage. Temperature conditioning, a process of exposure at a higher than normal recommended storage temperature for a given length of time, has been found successful in reducing chilling injury when grapefruit are subsequently subjected to the cold treatment (Hatton & Cubbedge 1983). Moreover, subjecting tomatoes (Klein & Lurie 1991) and mangoes (McCollum et al. 1993) to high temperatures (36 - 40°C) for 2 or 3 d prevented chilling injury in these commodities when they were held at temperatures required for quarantine security.

Conditioning commodities at high temperatures often allows normal softening and prevents damage to fruit following a heat treatment. The disruption of papaya softening by a quarantine heat treatment was reduced by a 4 h pretreatment at 42°C or a 1 h pretreatment at temperatures higher than 35°C followed by 3 h at 22°C (Paull & Chen 1990). Seo et al. (1974) reported an 8 h approach time to 44°C prevented subsequent injury when papaya were exposed to vapor heat. Conditioning 'Valencia' oranges and grapefruit for 8 h at 43°C allowed the fruit to tolerate subsequent exposure to a heat treatment (Sinclair & Lindgren 1955). Cucumbers conditioned by a 24 h treatment at 32.5°C were not damaged by a subsequent heat treatment (Chan & Linse 1989).

Decay was reduced in some fruit as a result of quarantine heat treatments. Brown et al. (1991) found that decay was partially controlled by a vapor heat treatment in grapefruit artificially inoculated with green-mold rot, *Penicillium digitatum* Sacc. Miller & McDonald (1991) found that decay was reduced in grapefruit as a result of a vapor heat treatment. Sharp & Spalding (1984) found immersion of mangoes in 46.1°C water reduced decay. McGuire (1991a) reported a reduction in decay of mangoes after immersion of the fruit in water at 46°C.

Irradiation

Gamma rays from a cobalt 60 source were applied to egg and larval infestations of oriental fruit fly, melon fly, and Mediterranean fruit fly in several Hawaii-grown fruits and vegetables to determine fly mortality and the effect on quality (Balock et al. 1966). Ten Hawaii-grown cultivars of avocados were extremely sensitive to 250 Gray (Gy), and 'Williams' bananas suffered skin russeting at 500 Gy. 'Pirie' mangoes were not adversely affected at doses up to 1,000 Gy, whereas 'Haden' mangoes tolerated a maximum of only 150 Gy without injury. Ten tomato cultivars tolerated doses of 250 to 1,000 Gy, and 'Solo' papayas showed no deleterious effects from doses up to 1,000 Gy. Four cultivars of lychee and cucumber tolerated doses of 500 Gy.

Decay of 'Tommy Atkins' mangoes ranged from 45% with 1,150 Gy using cesium 137 to 77% for fruit exposed to 750 Gy using cobalt 60 (Burditt et al. 1981). Irradiation caused dark, sunken areas in the surface of the skin of 'Keitt' mangoes, and the degree of injury was directly related to the irradiation dose. The lowest dose tested, 250 Gy, caused surface damage on 15% of the fruit (Burditt et al. 1981). In tests designed to determine the threshold of injury, Spalding & von Windeguth (1988) determined that a dose of ≥250 Gy should be avoided to minimize injury of 'Tommy Atkins' and 'Keitt' mangoes. Mitchell et al. (1990) concluded that the level of irradiation sufficient for quarantine purposes of 'Kensington Pride' mangoes (600 Gy) and 'Five Star' red capsicums (300 Gy) had no significant effect on their carotene levels.

'Duncan' grapefruit, 'Pineapple,' 'Valencia,' and 'Temple' oranges all showed irradiation-induced peel damage when treated with 1,000, 2,000, and 3,000 Gy (Dennison et al. 1966). The severity of peel injury was greater as the irradiation doses increased or as the storage temperature and duration increased. However, the organoleptic attributes of the citrus fruits were no different from those of control fruit. Burditt et al. (1981) exposed 'Marsh' grapefruit to 250 and 500 Gy, X rays or cobalt 60, or 500 Gy using cesium 137 as quarantine treatments against Caribbean fruit fly and reported that all treated fruits had a higher incidence of scald and decay. Fresh fruit sections treated with X rays had detectable taste differences and beta-pinene was absent from fruit irradiated with 500 Gy (Burditt et al. 1981). To further define the nature and severity of injury from irradiation, early-, mid-, and late-season 'Marsh' and 'Ruby Red' grapefruits were irradiated with cesium 137 at doses of 150, 300, 600, and 900 Gy (Hatton et al. 1982b). In most cases, rind breakdown increased progressively with increased irradiation dose, especially in mid- and late-season fruit. Irradiation doses of 150 and 300 Gy were satisfactory because fruits were acceptable with only slight rind injury, but irradiation doses of 600 and 900 Gy were unsatisfactory because of excessive rind breakdown. To compare different irradiation doses and rates of a single dose on peel injury response, Lester & Wolfenbarger (1990) exposed 'Ruby Red' grapefruit to cobalt 60 at either 200, 250, 300, or 350 Gy for 1 min, or 250 Gy for 1, 2.5, 5, or 25 min. Percentage electrolyte leakage (a direct measure of injury to the cell membrane) of the peel flavedo tissue was directly related to the level of irradiation dose. Grapefruit peel receiving 250 Gy over 2.5 or 5 min had a significantly lower percentage of electrolyte leakage than fruits receiving 250 Gy in 1 min, demonstrating that damage from irradiation is proportional to the rate of irradiation dose. Like electrolyte leakage, production of phenolics, a biochemical response to irradiation damage, in the peel following 250 Gy in 1, 2.5, 5, or 25 min increased as the rate of irradiation dose increased. Lester & Wolfenbarger (1990) concluded that a lower level of irradiation should impart little damage to grapefruit peel tissue.

Results by Jessup (1990) indicated that cherries irradiated up to 300 Gy for Queensland fruit fly infestation tended to remain darker red than controls. Irradiation caused peduncle discoloration of 'American Bing' cherries, but Jessup (1990) concluded that the cultivar would still be marketable.

'Arkin' carambola phytotoxicity tests at doses ranging from 50 to 1,500 Gy indicated that no observable damage occurred at levels between 50 and 500 Gy (Gould & von Windeguth 1991).

Combination Treatments

When two treatments alone have not proved effective from a pest mortality and/or fruit quality standpoint, a combination of the two treatments has been considered. One early report of a combination treatment is by Moffitt (1971) for codling moth disinfestation of 'Red Delicious' and 'Golden Delicious' apples. In this study, fumigation with 32 g per cubic meter of methyl bromide for 2 h at 23.9 - 25.6°C followed by a minimum storage period of 60 d at 0.6°C eliminated codling moth and did not reduce fruit quality or condition. However, unacceptable damage consisting of scald, accentuation of bruises, and accelerated breakdown in storage resulted when methyl bromide rates of 48 and 64 g per cubic meter were used.

Hayes et al. (1984) considered a thermal treatment against oriental fruit fly, consisting of exposing papayas to microwave radiation until a core temperature of 38 - 45°C was reached followed by a 20 min immersion in 48.7°C water and a 20 min immersion in 24°C water. About 17% of the fruit was damaged.

Limited observations of 'Marsh' grapefruit treated with gamma irradiation followed by cold storage against Caribbean fruit fly indicated no obvious visible damage to the fruit (von Windeguth & Gould 1990).

Jessup et al. (1988) studied the combination of dipping 'Kensington' mangoes in a benomyl solution (500 parts per million) or in water at 52°C for 5 min, followed by irradiation at doses up to 1,000 Gy, to provide quarantine security against Queensland fruit fly and mango weevil, *Cryptorhynchus mangiferae* (F.). Irradiation retarded skin color and increased skin wrinkling and skin bronzing with the effect being dose dependent. Significantly more skin bronzing and wrinkling occurred when mangoes were treated with heated water and heated benomyl. Dipped mangoes were found to be significantly softer than nondipped fruit.

Procedure for Typical Commodity Evaluation

Individual commodities will respond differently to the physical and chemical stresses imposed by quarantine treatments. Deleterious commodity responses may be conspicuous and appear immediately following treatment or only will become apparent after a storage or marketing period and include: abnormal weight loss, softening of pulp tissue, discoloration or scald of peel, lesions (pitting) of surface tissue, and increased incidence of rots. Some responses will be physiological in nature occurring especially in those commodities that are mature at the time of treatment but ripen after treatment. Symptoms of the latter responses include atypical peel or pulp color development, nonuniform softening of pulp, abnormal development of pulp texture, and off-flavors in juice or pulp.

Developing a comprehensive and useful scoring scheme that subjectively describes the characteristics of product condition based on physical and physiological parameters is important. The parameters for each characteristic should be well defined and documented both in the text and in photographs so examiners can systematically and accurately describe changes. Treated and nontreated control fruits should be examined before and after treatment and after storage and/or simulated marketing conditions. Some characteristics are objectively or subjectively measured while others are determined both objectively and subjectively. The rating scheme must provide for both meaningful and measurable resolutions of specific characteristics and for changes in those characteristics being examined. In general, the results of an effective rating scheme should furnish lucid indicators that will provide sound conclusions to assist in developing recommendations. Regulatory agencies and industry need to know not only the potential efficacy of the treatment, but also the effect of treatments on the condition and quality of commodities. Treatments should not be recommended that damage commodities or that cause downgrading that will affect marketing of those commodities. Three worksheet examples which were developed and are used for documenting changes in quality and condition characteristics of mango, carambola and grapefruit are described below.

Mango

Weight loss. Calculated from weight difference after treatment preparation and after a predetermined storage duration.

Maturity. Is the fruit mature, well developed? Rated qualitatively.

Subjective firmness. Rating based on yield to moderately applied finger pressure (MAFP): 1 = (hard), no yield to MAFP; 2 = (fairly hard), slight yield to MAFP; 3 = (fairly soft), moderate resistance to MAFP; 4 = (soft ripe), eating stage, slight resistance to MAFP; and 5 = (over ripe), mushy, little resistance to MAFP.

Peel color. 1 = 100% green; 2 = ≥1 ≤25% of ground color (yellow or red showing as green disappears); 3 = ≥26 ≤50% ground color; 4 = ≥51 ≤75% ground color; and 5 = ≥76% ground color.

Stage of ripeness. Based on a combination of subjective firmness and peel color rating: 1 = hard, 100% green; 2 = fairly hard, color of 2 or higher; 3 = fairly soft, firmness of 3 and color of 4 or 5; 4 = soft ripe, firmness of 4 and color of 4 or 5; and 5 = over ripe, firmness of 1 and color of 4 or 5. (Note: immature fruit may soften but remain green.)

Stem-end rot, **Diplodia natalensis** *P. Evans.* 1 = no decay, 2 = (trace), ≤3 mm spread from stem scar (petiole); 3 = (slight), ≥4 ≤13 mm; 4 = (moderate), ≥14 ≤25 mm; and 5 = (severe), ≥26 mm.

Anthracnose, **Colletotrichum gloeosporioides** *Penz.* 1 = no decay; 2 = (trace), ≤2% or less of aggregate surface area affected; 3 = (slight), ≥3 ≤10%; 4 = (moderate), ≥11 ≤20%; and 5 = (severe), ≥21%.

Other decay. If detected, score 1 = no decay; 2 = trace; 3 = slight; 4 = moderate; and 5 = severe.

Peel scald. Peel discoloration, may develop as brown or gray mottled areas or blush to random areas of peel tissue which may or may not had previous injury. 1 = no scald; 2 = ≤2% of surface area discolored; 3 = ≥3 ≤10%; 4 = ≥11 ≤25%; and 5 = ≥26% of surface area affected.

Pitting. Usually develops at random sites on fruit surface as individual or groups of stippled lesions which may or may not turn dark in color, or may affect small random areas (sheet pitting, contiguous lesions) of peel. The rating for pitting is the same as for peel scald.

Days to soft ripe. Designates the days for fruit to reach ripeness stage 4 after fruit are placed at 12°C.

Pulp discoloration. Observations of discoloration of vascular bundles and pulp tissue after slicing off a cheek and making a secondary transverse cut. Score for presence or absence and describe.

Remarks. Make notations of consistent trends such as symptoms of shriveling or other changes in condition characteristics that may relate to treatment.

Objective measurements. Pulp color is measured with a colorimeter. The pulp of each cheek is exposed by making a flat cut parallel to orientation of the seed and three color readings are made of each cheek and averaged. Pulp firmness is measured at one point per fruit with a firmness meter calibrated to read 10 N full scale, elongation of 5 mm and crosshead travel speed of 5 cm per min, with a 11.2 mm round penetrometer. Fruit quality assays include total soluble solids, pH, and acidity expressed as percent anhydrous citric acid.

Fruit number for ratings. For subjective ratings, usually 24 fruits per replication are examined, and three harvests are used as replications. For objective measurements, 24 fruits (held separately) per replication are used.

Frequency of inspections. Mangoes are initially selected before treatment to remove damaged and immature fruit and to obtain a uniform stage of ripeness and size depending upon the objectives of the test. Mangoes are then inspected after the quarantine treatment has been applied, after one and two weeks of storage at 12°C, and then holding at 21°C until they reach the soft-ripe stage.

Organoleptic evaluation. Flavor and texture are evaluated by an informal nine member panel. Individual members are aware of characteristic flavor and texture for a given commodity. Relative preference is indicated on a hedonic scale within a range from 0 to 100 or extremely unacceptable to extremely acceptable, respectively. Flavor considerations include the perception of sweet or sour, characteristic flavor and presence of off-flavors. If off-flavors are detected, then assays should be conducted in subsequent experiments for such things as ethanol, acetaldehyde, etc. Texture preference is based on masticatory sensations in response to the degree of pulp turgidity or crispness.

Flavor and texture preference should be determined on nonclimateric commodities (e.g. carambola, grapefruit) before treatment as well as following treatment and storage. Climacteric commodities (e.g. mango) are only evaluated at the edible stage of ripeness.

Carambola

Weight loss and maturity. See mango.

Subjective firmness. Rating based on yield to MAFP and resistance to twist (hold each end of carambola and apply slight twist to ends of fruit in opposite direction). 1 = (flaccid), little resistance to force; 2 (fairly firm), moderate resistance to force; and 3 = (firm), resistant to force.

Peel color. 1 = 100% green; 2 = breaker, slight yellow blush, (≤3% of surface area yellow); 3 = ≥4 - ≤75% yellow; 4 = green on outer edges of ribs only and ≥76% yellow; and 5 = 100% yellow/orange.

Stem-end rot. Score for presence or absence.

Miscellaneous decay. Score for presence or absence.

Total decay. Total of stem-end rot and miscellaneous decay although decay is rarely observed.

Superficial mold. May develop on sites of senescent tissue or at sites of injury. Score for presence or absence.

Pitting. Usually develops at random sites on fruit surface as individual or groups of stippled lesions which may or may not become discolored. 1 = no injury; 2 = ≤10% of surface area; 3 = ≥11 - ≤25%; 4 = ≥26 - ≤50%; and 5 = ≥51% of fruit surface area affected.

Scald. A discoloration of the fruit surface and is often referred to as bronzing. Scored the same as pitting.

Stem-end breakdown. Tissue at the stem-end becomes senescent and turns dark-brown. Usually occurs after shriveling develops and is scored the same as pitting.

Fin browning. Tissue at outer edges of ribs becomes senescent and turns brown. Scoring is based on the degree of severity and amount of tissue affected: 1 = no injury; 2 = slight senescent area on one rib edge; 3 = moderately senescent areas or two or more rib edges, affected area extending ≤3 mm perpendicular from rib edge; 4 = damage extends ≥ 3 mm perpendicular from rib edge on one rib; and 5 = damage extends ≥3 mm perpendicular from rib edge on more than one rib.

Pulp discoloration. Fruit are transverse cut and presence of any internal discoloration is recorded.

Remarks. Any abnormality is noted that may be related to treatment.

Injury. During initial inspection, the presence of any mechanical injury is noted.

Objective measurements. Peel color is measured with a colorimeter and is determined on three predetermined and marked sites on each fruit. Pulp firmness is measured with a firmness meter calibrated to read 10 N full scale, elongation of 3 mm and crosshead travel speed of 5 cm per min, with a 11.2 mm round penetrometer. Individual fruit are set on a support jig and firmness on three ribs per fruit is determined and averaged. Fruit quality assays include total soluble solids, pH, and acidity expressed as percent anhydrous oxalic acid.

Fruit number for ratings. For subjective ratings, usually 60 fruits per replication are examined, and three harvests are used as replications. For objective measurements, 15 fruits (held separately) per replication are used.

Frequency of inspections. Carambolas are initially selected before treatment to remove damaged and immature fruit and to obtain a uniform stage of ripeness and size depending upon the objectives of the test. Carambolas are then inspected after the quarantine treatment has been applied, after one week of storage at 5°C, and after holding 3 d at 15°C.

Organoleptic evaluation. See mango.

Grapefruit

Count. Number of fruit per 4/5 bushel standard citrus box.

Sound fruit. Number of damage-free fruit per box.

Aging. Symptoms of senescence (rind breakdown, necrotic tissue) at the stem end of fruit. Record severity and number of fruit affected. 1 = no injury; 2 = (slight), ≤6 mm diameter aggregate area; 3 = (moderate), ≥7 - ≤12 mm aggregate diameter area; and 4 = (severe), ≥13 mm diameter aggregate area.

Pitting. Abruptly sunken spots developing randomly on the peel that are not discolored at first, but later become tan to brown. Scored by severity as percentage of surface area affected. 1 = no injury; 2 = ≤10%; 3 = ≥11 - ≤25%; 4 = ≥26 - ≤50%; and 5 = ≥51% of surface area affected.

Scald. Tan to brown discoloration of the peel. Patches of discoloration may develop on peel tissue due to fruit response to treatment stress and is scored the same as pitting.

Other. Define any other unexpected disorder that may appear, and quantify by developing a rating scheme.

Blue-mold, Penicillium italicum *Wehmer, and green-mold rots,* P. digitatum *Sacc..* Distinguish and record occurrence of each.

Stem-end rots, Phomopsis citri *Fawc. and* Diplodia natalensis *P. Evans.* Distinguish and record occurrence of each.

Miscellaneous decay. Includes alternaria rot, *Alternaria citri* Ell. & Pierce, anthracnose, and sour rot, *Geotrichum candidum* Lk. ex Pers. Distinguish and record occurrence of each.

Total decay. Cumulative number of fruit with decay after designated storage duration.

Fruit appearance. Based on freshness appearance of peel. 1 = fresh and glossy; 2 = fairly fresh, some loss of gloss; and 3 = old, dull, generally without gloss. Rating is a single value based on an entire carton of fruit.

Stem condition. Score given when calyx is attached (generally detached in degreened fruit) and based on color of calyx stem button. 1 = green; 2 = green/brown; and 3 = brown. Rating is a single value based on an entire carton of fruit.

Fruit firmness. Subjective rating. 1 = firm, slight yield MAFP; 2 = fairly firm, moderate yield to MAFP; and 3 = soft, little resistance to MAFP. Rating is a single value based on an entire carton of fruit.

Fruit color. Subjective rating. 1 = good color, fresh-yellow hue, may have some green blush; 2 = fair, yellow, but moderately fresh to dull-yellow hue; and 3 = poor, darker yellow hue, dull sheen without wax and with orange blush.

Rind condition. Listing made of preharvest or pretreatment conditions such as sunburn, wind scaring, or oleocellosis at initial fruit evaluation.

Comments. All abnormal conditions not categorized are noted during the course of fruit evaluations.

Objective measurements. Peel color is measured with a colorimeter and is determined at three locations around the equator of a fruit. Fruit firmness is determined at one point on the equator of a fruit with a firmness meter calibrated to read 10 N full scale, elongation of 5 mm and crosshead travel speed of 5 cm per min, with a 5.7 cm diameter flat surface anvil. Fruit quality assays include total soluble solids, pH, and acidity expressed as percent anhydrous citric acid.

Fruit number for ratings. For subjective ratings, usually 120 fruits per replication are examined, and three harvests are used as replications. For objective measurements, 15 fruits (held separately) per replication are used.

Frequency of inspections. Grapefruit is selected before treatment to remove damaged, misshaped, and off-bloom fruit. Fruit of a uniform size is selected depending upon the objective of the test. Grapefruit is then inspected after the quarantine treatment has been applied, after two and four weeks storage at 15°C, and after holding one week at 21°C. After January 1, fruits are stored at 10°C instead of 15°C.

Organoleptic evaluation. See mango.

From the inspection procedures for these three different fruits, it is obvious that a single procedure can not be used for all commodities. When developing an inspection procedure that considers the influence of a quarantine treatment on the condition and quality of a particular commodity, one must be familiar with those attributes that are distinctive to that commodity. In this way one can determine the effect, if any, a quarantine treatment under development will have on the condition and quality of a commodity.

Conclusions

Several chemical and physical measures are effective as treatments against quarantined insects. Some treatments have no deleterious effects on the condition and quality of some commodities and cultivars. However, injury can occur at times with approved treatments under commercial conditions.

Investigations are continuing to determine the influence of factors such as maturity, time after harvest, and post treatment storage conditions on susceptibility to injury. Research seeks to further refine the currently approved treatments and to develop alternatives. Because consumers have become increasingly cautious about chemical treatments, substantial interest exists in physical treatments, especially the use of temperature management.

It is important that treatments developed under laboratory conditions be feasible in a commercial setting. The treatment protocol must accommodate not only the variability in commodity condition, but also treatment variations which occur under commercial conditions without leading to commodity damage or insect survival.

TABLE 17.1   Response of fruits and vegetables to quarantine treatment with respect to damage

Commodity	Treatment	Damage	Reference
Apple	Combination	No	Moffitt (1971)
Apple	Fumigation	Yes	Drake et al. (1988)
Apple	Fumigation	Yes	Kenworthy (1944)
Apple	Fumigation	Yes	Kenworthy & Gaddis (1946)
Avocado	Fumigation	Yes	Lindgren & Sinclair (1951)
Avocado	Irradiation	Yes	Balock et al. (1966)
Avocado	Vapor heat	Yes	Sinclair & Lindgren (1955)
Banana	Hot water	No	Armstrong (1982)
Banana	Irradiation	No	Balock et al. (1966)
Carambola	Cold	No	Gould & Sharp (1990b)
Carambola	Cold	No	Miller et al. (1991c)
Carambola	Fumigation	Yes	Hallman & King (1992)
Carambola	Heated forced air	Yes	Miller et al. (1990)
Carambola	Hot water	Yes	Hallman (1989)
Carambola	Hot water	Yes	Hallman (1991)
Carambola	Hot water	Yes	Hallman & Sharp (1990)
Carambola	Irradiation	No	Gould & von Windeguth (1991)
Carambola	Vapor heat	No	Hallman (1990)
Carambola	Vapor heat	Yes	Hallman (1991)
Cherry	Fumigation	No	Anthon et al. (1975)
Cherry, sweet	Irradiation	No	Jessup (1990)
Citrus	Vapor heat	No	Baker (1952)

(Continues)

TABLE 17.1. (*Continued*)

Commodity	Treatment	Damage	Reference
Cucumber	Fumigation	No	Armstrong & Garcia (1985)
Cucumber	Hot water	No	Chan & Linse (1989)
Cucumber	Irradiation	No	Balock et al. (1966)
Deciduous fruits	Fumigation	Yes	Claypool & Vines (1956)
Eggplant	Vapor heat	No	Furusawa et al. (1984)
Grapefruit	Cold	No	Chalutz et al. (1985)
Grapefruit	Cold	No	Hatton & Cubbedge (1982)
Grapefruit	Cold	No	Hatton & Cubbedge (1983)
Grapefruit	Combination	No	von Windeguth & Gould (1990)
Grapefruit	Film packaging	No	Sharp (1990b)
Grapefruit	Fumigation	Yes	Hatton et al. (1982a)
Grapefruit	Fumigation	Yes	Lindgren & Sinclair (1951)
Grapefruit	Heated forced air	No	McGuire (1991b)
Grapefruit	Heated forced air	No	Sharp (1989)
Grapefruit	Hot water	Yes	McGuire (1991b)
Grapefruit	Hot water	Yes	Miller et al. (1988)
Grapefruit	Hot water	Yes	Miller et al. (1989)
Grapefruit	Hot water	Yes	Sharp (1985)
Grapefruit	Irradiation	Yes	Dennison et al. (1966)
Grapefruit	Irradiation	No	Hatton et al. (1982b)
Grapefruit	Irradiation	No	Lester & Wolfenbarger (1990)
Grapefruit	Vapor heat	No	Hallman et al. (1990)
Grapefruit	Vapor heat	No	Miller & McDonald (1991)
Grapefruit	Vapor heat	No	Miller & McDonald (1992)
Grapefruit	Vapor heat	Yes	Miller et al. (1989)
Grapefruit	Vapor heat	No	Miller et al. (1991b)
Grapefruit	Vapor heat	Yes	Sinclair & Lindgren (1955)
Guava	Hot water	No	Gould & Sharp (1992)
Guava	Vapor heat	No	Sein (1935)
Lemon	Cold	No	Houck et al. (1990)
Lemon	Fumigation	Yes	Lindgren & Sinclair (1951)
Lemon	Vapor heat	Yes	Sinclair & Lindgren (1955)
Lychee	Irradiation	No	Balock et al. (1966)
Mango	Combination	Yes	Jessup et al. (1988)
Mango	Film packaging	Yes	Gould & Sharp (1990a)
Mango	Heated forced air	No	McGuire (1991a)
Mango	Heated forced air	No	Miller et al. (1991a)
Mango	Hot water	Yes	Jacobi & Wang (1991)
Mango	Hot water	No	Segarra-Carmona et al.(1990)
Mango	Hot water	No	Sharp (1986)
Mango	Hot water	No	Sharp et al. (1989a)
Mango	Hot water	No	Sharp et al. (1989b)
Mango	Hot water	No	Sharp et al. (1989c)
Mango	Hot water	No	Sharp et al. (1988)
Mango	Hot water	No	Sharp & Spalding (1984)

(Continues)

TABLE 17.1   (*Continued*)

Commodity	Treatment	Damage	Reference
Mango	Hot water	No	Spalding et al. (1988)
Mango	Irradiation	Yes	Balock et al. (1966)
Mango	Irradiation	Yes	Burditt et al. (1981)
Mango	Irradiation	No	Mitchell et al. (1990)
Mango	Irradiation	Yes	Spalding & von Windeguth (1988)
Mango	Vapor heat	No	Sein (1935)
Mango	Vapor heat	No	Sunagawa et al. (1987)
Nectarine	Fumigation	No	Harvey et al. (1989)
Nectarine	Fumigation	No	Yokoyama et al. (1990)
Nectarine	Hot water	Yes	Sharp (1990a)
Orange	Fumigation	Yes	Lindgren & Sinclair (1951)
Orange	Irradiation	Yes	Dennison et al. (1966)
Orange	Vapor heat	Yes	Sinclair & Lindgren (1955)
Papaya	Combination	Yes	Hayes et al. (1984)
Papaya	Heated forced air	No	Armstrong et al. (1989)
Papaya	Hot water	No	Couey & Hayes (1986)
Papaya	Irradiation	No	Balock et al. (1966)
Papaya	Vapor heat	No	Jones et al. (1939)
Peach	Hot water	Yes	Sharp (1990a)
Pepper, green	Vapor heat	No	Sugimoto et al. (1983)
Pepper, red	Irradiation	No	Mitchell et al. (1990)
Plum	Hot water	Yes	Sharp (1990a)
Tomato	Irradiation	No	Balock et al. (1966)

References

Anonymous. 1991. Department of Agriculture, Animal and Plant Health Inspection Service. Papayas from Hawaii, final rule. Federal Register 56: 59205-59207.

Anthon, E. W., H. R. Moffitt, H. M. Couey & L. O. Smith. 1975. Control of codling moth in harvested sweet cherries with methyl bromide and effects upon quality and taste of treated fruit. J. Econ. Entomol. 68: 524-526.

Armstrong, J. W. 1982. Development of a hot-water immersion quarantine treatment for Hawaiian-grown 'Brazilian' bananas. J. Econ. Entomol. 75: 787-790.

Armstrong, J. W. & D. L. Garcia. 1985. Methyl bromide quarantine fumigations for Hawaii-grown cucumbers infested with melon fly and oriental fruit fly (Diptera: Tephritidae). J. Econ. Entomol. 78: 1308-1310.

Armstrong, J. W., J. D. Hansen, B.K.S. Hu & S. A. Brown. 1989. High-temperature, forced-air quarantine treatment for papayas infested with Tephritid fruit flies (Diptera: Tephritidae). J. Econ. Entomol. 82: 1667-1674.

Baker, A. C. 1952. "The Vapor-Heat Process," in *Insects: The Yearbook of Agriculture*. Pp. 401-404. Department of Agr. U.S. Government Printing Office. Washington, D.C.

Balock, J. W., A. K. Burditt, Jr., S. T. Seo & E. K. Akamine. 1966. Gamma radiation as a quarantine treatment for Hawaiian fruit flies. J. Econ. Entomol. 59: 202-204.

Brown, G. E., W. R. Miller & R. E. McDonald. 1991. Control of green mold in 'Marsh' grapefruit with vapor heat quarantine treatment. Proceedings, Fla. State Hortic. Soc. 104: 115-117.

Burditt, A. K., Jr., M. G. Moshonas, T. T. Hatton, D. H. Spalding, D. L. von Windeguth & P. E. Shaw. 1981. Low-dose irradiation as a treatment for grapefruit and mangoes infested with Caribbean fruit fly larvae. USDA, Agr. Research Service Report ARR-S-10. Pp. iii+9.

Chalutz, E., J. Waks & M. Schiffmann-Nadel. 1985. Reducing susceptibility of grapefruit to chilling injury during cold treatment. HortScience 20: 226-228.

Chan, H. T., Jr. 1988. Effects of heat treatments on the ethylene forming enzyme system in papayas. J. Food Sci. 51: 581-583.

Chan, H. T., Jr., & E. Linse. 1989. Conditioning cucumbers for quarantine heat treatments. HortScience 24: 985-989.

Chan, H. T., Jr., S.Y.T. Tam & S. T. Seo. 1981. Papaya polygalacturonase and its role in thermally injured ripening fruit. J. Food Sci. 46: 190-191, 197.

Claypool, L. L. & H. M. Vines. 1956. Commodity tolerance studies of deciduous fruits to moist heat and fumigants. Hilgardia 24: 297-355.

Couey, H. M. & C. F. Hayes. 1986. Quarantine procedure for Hawaiian papaya using fruit selection and a two-stage hot-water immersion. J. Econ. Entomol. 79: 1307-1314.

Dennison, R. A., W. Grierson & E. M. Ahmed. 1966. Irradiation of 'Duncan' grapefruit, 'Pineapple' and 'Valencia' oranges and 'Temples.' Proceedings, Fla. State Hortic. Soc. 79: 285-292.

Drake, S. R., H. R. Moffitt, J. K. Fellman & C. R. Sell. 1988. Apple quality as influenced by fumigation with methyl bromide. J. Food Sci. 53: 1710-1712, 1736.

Furusawa, K., T. Sugimoto & T. Gaja. 1984. The effectiveness of vapor heat treatment against the melon fly, *Dacus cucurbitae* Coquillett, in eggplant and fruit tolerance to the treatment. Research Bul. Plant Protection Service (Japan) 20: 17-24 (in Japanese).

Gould, W. P. & J. L. Sharp. 1990a. Caribbean fruit fly (Diptera: Tephritidae) mortality induced by shrink-wrapping infested mangoes. J. Econ. Entomol. 83: 2324-2326.

_____. 1990b. Cold-storage quarantine treatment for carambolas infested with the Caribbean fruit fly (Diptera: Tephritidae). J. Econ. Entomol. 83: 458-460.

_____. 1992. Hot-water immersion quarantine treatment for guavas infested with Caribbean fruit fly (Diptera: Tephritidae). J. Econ. Entomol. 85: 1235-1239.

Gould, W. P. & D. L. von Windeguth. 1991. Gamma irradiation as a quarantine treatment for carambolas infested with Caribbean fruit flies. Fla. Entomologist 74: 297-300.

Hallman, G. J. 1989. Quality of carambolas subjected to hot-water immersion quarantine treatment. Proceedings, Fla. State Hortic. Soc. 102: 155-156.

_____. 1990. Vapor-heat treatment of carambolas infested with Caribbean fruit fly (Diptera: Tephritidae). J. Econ. Entomol. 83: 2340-2342.

_____. 1991. Quality of carambolas subjected to postharvest hot water immersion and vapor heat treatments. HortScience 26: 286-287.

Hallman, G. J. & J. R. King. 1992. Methyl bromide fumigation quarantine treatment for carambolas infested with Caribbean fruit fly (Diptera: Tephritidae). J. Econ. Entomol. 85: 1231-1234.

Hallman, G. J. & J. L. Sharp. 1990. Hot-water immersion quarantine treatment for carambolas infested with Caribbean fruit fly (Diptera: Tephritidae). J. Econ. Entomol. 83: 1471-1474.

Hallman, G. J., J. J. Gaffney & J. L. Sharp. 1990. Vapor heat treatment for grapefruit infested with Caribbean fruit fly (Diptera: Tephritidae). J. Econ. Entomol. 83: 1475-1478.

Harvey, J. M., C. M. Harris & P. L. Hartsell. 1989. Tolerances of California nectarine cultivars to methyl bromide quarantine treatments. J. American Soc. Hortic. Sci. 114: 626-629.

Hatton, T. T. & R. H. Cubbedge. 1982. Conditioning Florida grapefruit to reduce chilling injury during low-temperature storage. J. American Soc. Hortic. Sci. 107: 57-60.

_____. 1983. Preferred temperature for prestorage conditioning of 'Marsh' grapefruit to prevent chilling injury at low temperatures. HortScience 18: 721-722.

Hatton, T. T., R. H. Cubbedge, D. L. von Windeguth & D. H. Spalding. 1982. Control of Caribbean fruit fly in Florida grapefruit by phosphine fumigation. Proceedings, Fla. State Hortic. Soc. 95: 221-224.

Hatton, T. T., R. H. Cubbedge, L. A. Risse, P. W. Hale, D. H. Spalding & W. F. Reeder. 1982. Phytotoxicity of gamma irradiation on Florida grapefruit. Proceedings, Fla. State Hortic. Soc. 95: 232-234.

Hayes, C. F., H.T.G. Chingon, F. A. Nitta & W. J. Wang. 1984. Temperature control as an alternative to ethylene dibromide fumigation for the control of fruit flies (Diptera: Tephritidae) in papaya. J. Econ. Entomol. 77: 683-686.

Houck, L. G., J. F. Jenner & B. E. Mackey. 1990. Seasonal variability of the response of desert lemons to rind injury and decay caused by quarantine cold treatments. J. Hortic. Sci. 65: 611-617.

Jacobi, K. R. & L.S. Wong. 1991. The injuries and changes in ripening behaviour caused to Kensington mango by hot water treatment. Acta Horticulturae 291: 372-378.

Jessup, A. J. 1990. Gamma irradiation as a quarantine treatment for sweet cherries against Queensland fruit fly. HortScience 25: 456-458.

Jessup, A. J., C. J. Rigney & P. A. Wills. 1988. Effects of gamma irradiation combined with hot dipping on quality of 'Kensington Pride' mangoes. J. Food Sci. 53: 1486-1489.

Jones, W. W. 1940. Vapor-heat treatment for fruits and vegetables grown in Hawaii. Hawaii Agric. Experimental Station Circular 16.

Jones, W. W., J. J. Holzman & A. G. Galloway. 1939. The effect of high-temperature sterilization on the Solo papaya. Hawaii Agric. Experiment Station Circular 14.

Kenworthy, A. L. 1944. Injury to 'Williams' apples resulting from fumigation with methyl bromide. Proceedings, American Soc. Hortic. Sci. 45: 141-145.

Kenworthy, A. L. & C. H. Gaddis. 1946. Tolerance of apple varieties to methyl bromide fumigation. Proceedings, American Soc. Hortic. Sci. 47: 64-66.

Klein, J. D. & S. Lurie. 1991. Prestorage heat stress protects tomatoes against chilling injuries. Hassadeh 71:542-544 (in Hebrew).

Lester, G. E. & D. A. Wolfenbarger. 1990. Comparisons of cobalt-60 gamma irradiation dose rates on grapefruit flavedo tissue and on Mexican fruit fly mortality. J. Food Protection 53: 329-331.

Lindgren, D. L. & W. B. Sinclair. 1951. Tolerance of citrus and avocado fruits to fumigants effective against the Oriental fruit fly. J. Econ. Entomol. 44: 980-990.

McCollum, T. G., S. D'Aquino & R. E. McDonald. 1993. Heat treatment inhibits mango chilling injury. HortScience 28: 197-198.

McGuire, R. G. 1991a. Concomitant decay reductions when mangos are treated with heat to control infestations of Caribbean fruit flies. Plant Disease 75: 946-949.

_____. 1991b. Market quality of grapefruit after heat quarantine treatments. HortScience 26: 1393-1395.

Miller, W. R. & R. E. McDonald. 1991. Quality of stored 'Marsh' and 'Ruby Red' grapefruit after high-temperature, forced-air treatment. HortScience 26: 1188-1991.

_____. 1992. Postharvest quality of early season grapefruit after forced-air vapor heat treatment. HortScience 27: 422-424.

Miller, W. R., R. E. McDonald, T. T. Hatton & M. Ismail. 1988. Phytotoxicity to grapefruit exposed to hot water immersion treatment. Proceedings, Fla. State Hortic. Soc. 101: 192-195.

Miller, W. R., R. E. McDonald, G. J. Hallman & M. Ismail. 1989. Phytotoxicity of hot water and vapor heat treatments to Florida grapefruit. Proceedings, International Conference on Technical Innovation in Freezing and Refrigeration of Fruits and Vegetables. Commissions C2, D1, D2, and D3. Pp. 207-212. University of California at Davis.

Miller, W. R., R. E. McDonald & J. L. Sharp. 1990. Condition of Florida carambolas after hot-air treatment and storage. Proceedings, Fla. State Hortic. Soc. 103: 238-241.

_____. 1991a. Quality changes during storage and ripening of 'Tommy Atkins' mangos treated with heated forced air. HortScience 26: 395-397.

Miller, W. R., R. E. McDonald, G. J. Hallman & J. L. Sharp. 1991b. Condition of Florida grapefruit after exposure to vapor heat quarantine treatment. HortScience 26: 42-44.

Miller, W. R., R. E. McDonald & M. Nisperos-Carriedo. 1991c. Quality of 'Arkin' carambolas with or without conditioning followed by low-temperature quarantine treatment. Proceedings, Fla. State Hortic. Soc. 104: 118-122.

Mitchell, G. E., R. L. McLauchlan, T. R. Beattie, C. Banos & A. A. Gillen. 1990. Effect of gamma irradiation on the carotene content of mangos and red capsicums. J. Food Sci. 55: 1185-1186.

Moffitt, H. R. 1971. Methyl bromide fumigation combined with storage for control of codling moth in apples. J. Econ. Entomol. 64: 1258-1260.

Paull, R. E. & N. J. Chen. 1990. Heat shock response in field-grown, ripening papaya fruit. J. American Soc. Hortic. Sci. 115: 623-631.

Ruckelshaus, W. D. 1984. Ethylene dibromide, amendment of notice of intent to cancel registration of pesticide products containing ethylene dibromide. Federal Register 49: 14182-14185.

Segarra-Carmona, A. E., R. A. Franqui, L. V. Ramirez-Ramos, L. R. Santiago & C. N. Torres-Rivera. 1990. Hot-water dip treatments to destroy *Anastrepha obliqua* larvae (Diptera: Tephritidae) in mangoes from Puerto Rico. J. Agr. University of Puerto Rico. 74: 441-447.

Sein, F., Jr. 1935. Heat sterilization of mangoes and guavas for fruit flies. J. Agr. University of Puerto Rico. 19: 105-115.

Seo, S. T., B.K.S. Hu, M. Komura, C.Y.L. Lee & E. J. Harris. 1974. *Dacus dorsalis*: Vapor heat treatment in papaya. J. Econ. Entomol. 67: 240-242.

Sharp, J. L. 1985. Submersion of Florida grapefruit in heated water to kill stages of Caribbean fruit fly, *Anastrepha suspensa*. Proceedings, Fla.

State Hortic. Soc. 98: 78-80.

_____. 1986. Hot-water treatment for control of *Anastrepha suspensa* (Diptera: Tephritidae) in mangos. J. Econ. Entomol. 79: 706-708.

_____. 1989. Preliminary investigation using hot air to disinfest grapefruit of Caribbean fruit fly immatures. Proceedings, Fla. State Hortic. Soc. 102: 157-159.

_____. 1990a. Immersion in heated water as a quarantine treatment for California stone fruits infested with the Caribbean fruit fly (Diptera: Tephritidae). J. Econ. Entomol. 83: 1468-1470.

_____. 1990b. Mortality of Caribbean fruit fly immatures in shrinkwrapped grapefruit. Fla. Entomologist 73: 660-665.

Sharp, J. L. & D. H. Spalding. 1984. Hot water as a quarantine treatment for Florida mangos infested with Caribbean fruit fly. Proceedings, Fla. State Hortic. Soc. 97: 355-357.

Sharp, J. L., M. T. Ouye, R. Thalman, W. Hart, S. Ingle & V. Chew. 1988. Submersion of 'Francis' mango in hot water as a quarantine treatment for the West Indian fruit fly and the Caribbean fruit fly (Diptera: Tephritidae). J. Econ. Entomol. 81: 1431-1436.

Sharp, J. L., M. T. Ouye, S. J. Ingle & W. G. Hart. 1989a. Hot-water quarantine treatment for mangoes from Mexico infested with Mexican fruit fly and West Indian fruit fly (Diptera: Tephritidae). J. Econ. Entomol. 82: 1657-1662.

Sharp, J. L., M. T. Ouye, W. Hart, S. Ingle, G. J. Hallman, W. Gould & V. Chew. 1989b. Immersion of Florida mangos in hot water as a quarantine treatment for Caribbean fruit fly (Diptera: Tephritidae). J. Econ. Entomol. 82: 186-188.

Sharp, J. L., M. T. Ouye, S. J. Ingle, W. G. Hart, W. R. Enkerlin H., H. Celedonio H., J. Toledo A., L. Stevens, E. Quintero, J. Reyes F. & A. Schwarz. 1989c. Hot-water quarantine treatment for mangoes from the state of Chiapas, Mexico, infested with Mediterranean fruit fly and *Anastrepha serpentina* (Wiedemann) (Diptera: Tephritidae). J. Econ. Entomol. 82: 1663-1666.

Sinclair, W. B. & D. L. Lindgren. 1955. Vapor heat sterilization of California citrus and avocado fruits against fruit-fly insects. J. Econ. Entomol. 48: 133-138.

Spalding, D. H. & D. L. von Windeguth. 1988. Quality and decay of irradiated mangos. HortScience 23: 187-189.

Spalding, D. H., J. R. King & J. L. Sharp. 1988. Quality and decay of mangos treated with hot water for quarantine control of fruit fly. Tropical Sci. 28: 95-101.

Sugimoto, T., K. Furusawa & M. Mizobuchi. 1983. The effectiveness of vapor heat treatment against the oriental fruit fly, *Dacus dorsalis* Hendel, in green pepper and fruit tolerance to the treatment.

Research Bul. Plant Protection Service (Japan) 19: 81-88 (in Japanese).

Sunagawa, K., K. Kume & R. Iwaizumi. 1987. The effectiveness of vapor heat treatment against the melon fly, *Dacus cucurbitae* Coquillett, in mango and fruit tolerance to the treatment. Research Bul. Plant Protection Service (Japan) 23: 13-20 (in Japanese).

von Windeguth, D. L. & W. P. Gould. 1990. Gamma irradiation followed by cold storage as a quarantine treatment for Florida grapefruit infested with Caribbean fruit fly. Fla. Entomologist 73: 242-247.

Yokoyama, V. Y., G. T. Miller & P. L. Hartsell. 1990. A methyl bromide quarantine treatment to control codling moth (Lepidoptera: Tortricidae) on nectarines packed in shipping containers for export to Japan and effect on fruit attributes. J. Econ. Entomol. 83: 2335-2339.

About the Contributors

John W. Armstrong, Ph.D., is a Research Entomologist at the Tropical Fruit & Vegetable Research Laboratory, Agricultural Research Service, U.S. Department of Agriculture, P.O. Box 4459, Hilo, HI 96720.

James R. Brazzel, Ph.D., is Director of Mission Plant Methods Center, Route 3, Box 1000, Edinburg, TX 78539.

Arthur K. Burditt, Jr., Ph.D., is a retired Research Entomologist formerly with the Agricultural Research Service, U.S. Department of Agriculture. He now resides at A590-C SW 82 Terrace, Ocala, FL 32676.

Alan Carpenter, Ph.D., is Team Leader of the Postharvest Disinfestation Programme at the New Zealand Institute for Crop & Food Research, Ltd., Ministry of Agriculture & Fisheries, Private Bag, Kimberly Road, Levin, New Zealand.

Victor Chew is a Statistician in the Office of the Area Director, South Atlantic Region, Agricultural Research Service, U.S. Department of Agriculture, located at the University of Florida, 412 Rolfs Hall, Gainesville, FL 32611.

E. Ruth Frampton, Ph.D., is a National Policy Advisor for the Ministry of Agriculture & Fisheries, Canterbury Agriculture & Science Center, Ellesmere Junction Road, P.O. Box 24, Lincoln, New Zealand.

Walter P. Gould, Ph.D., is a Research Entomologist at the Subtropical Horticulture Research Station, Agricultural Research Service, U.S. Department of Agriculture, 13601 Old Cutler Road, Miami, FL 33158.

Guy J. Hallman, Ph.D., is a Research Entomologist at the Subtropical Horticulture Research Station, Agricultural Research Service, U.S. Department of Agriculture, 13601 Old Cutler Road, Miami, FL 33158.

Neil W. Heather, Ph.D., is a Supervising Entomologist at the Queensland Department of Primary Industries, Meiers Road, Indooroopilly, QLD 4068, Australia.

Eric B. Jang, Ph.D., is Research Leader and a Research Entomologist at the Tropical Fruit & Vegetable Research Laboratory, Agricultural Research Service, U.S. Department of Agriculture, P.O. Box 4459, Hilo, HI 96720.

Robert L. Mangan, Ph.D., is Research Leader and a Research Entomologist for Crop Quality & Fruit Insects Research at the Subtropical Agriculture Research Laboratory, Agricultural Research Service, U.S. Department of Agriculture, 301 S. International Boulevard, Weslaco, TX 78596.

Roy E. McDonald, Ph.D., is Research Leader and a Research Horticulturist in Export and Quality Improvement Research at the U.S. Horticultural Research Laboratory, Agricultural Research Service, U.S. Department of Agriculture, 2120 Camden Road, Orlando, FL 32803.

William R. Miller is a Research Agricultural Marketing Specialist in Export and Quality Improvement Research at the U.S. Horticultural Research Laboratory, Agricultural Research Service, U.S. Department of Agriculture, 2120 Camden Road, Orlando, FL 32803.

Harold R. Moffitt, Ph.D., is a Research Entomologist in Fruit and Vegetable Insect Research at the Yakima Agricultural Research Laboratory, Agricultural Research Service, U.S. Department of Agriculture, 3706 W. Nob Hill Boulevard, Yakima, WA 98902.

Ru Nguyen, Ph.D., is a Biological Scientist with the Division of Plant Industry, Florida Department of Agriculture & Consumer Services, P.O. Box 147100, Gainesville, FL 32614-7100.

Murray Potter, Ph.D., formerly a Post-Doctorate Fellow at the Levin Research Centre, currently lectures in the Department of Ecology on entomology and wildlife ecology at Massey University, Palmerston North, New Zealand.

Haiganoush K. Preisler, Ph.D., is a Research Statistician with Dr. Jacqueline Robertson's project at the Pacific Southwest Forest & Range Experiment Station of the Forest Service, U.S. Department of Agriculture, 1960 Addison Street, Berkeley, CA 94701.

Connie Riherd is Assistant Director, Division of Plant Industry, Florida Department of Agriculture & Consumer Services, P.O. Box 147100, Gainesville, FL 32614-7100.

Jacqueline L. Robertson, Ph.D., is a Research Entomologist at the Pacific Southwest Forest & Range Experiment Station of the Forest Service, U.S. Department of Agriculture, 1960 Addison Street, Berkeley, CA 94701. She also serves as editor for the *Journal of Economic Entomology* and subject editor for *Canadian Entomologist.*

Michael J. Shannon is the Assistant Director of Operational Support in the Director's Office of Plant Protection and Quarantine, Animal and Plant Health Inspection Service, U.S. Department of Agriculture, 6505 Belcrest Road, Room 649-FB, Hyattsville, MD 20782.

Jennifer L. Sharp, Ph.D., is Research Leader, Location Coordinator, and a Research Entomologist at the Subtropical Horticulture Research Station, Agricultural Research Service, U.S. Department of Agriculture, 13601 Old Cutler Road, Miami, FL 33158.

Susan P. Worner, Ph.D., is a faculty member in the Department of Entomology & Animal Ecology at Lincoln University, P.O. Box 84, Canterbury, New Zealand.

Victoria Y. Yokoyama, Ph. D., is a Research Entomologist in Commodity Protection and Quarantine Research at the Horticultural Crops Research Laboratory, Agricultural Research Service, U.S. Department of Agriculture, 2021 S. Peach Avenue, Fresno, CA 93727.

Index

Acanthoscelides obtectus Say, 107
Acarus siro L., 180
Acceptable daily intake (ADI), 89-90
Acrylonitrile, 251
Aculus schlechtendali (Nalepa), 179
Adenosine triphosphate, 172
ADI (see Acceptable daily intake)
Aleurocanthus woglumi Ashby, 74
Alfalfa weevil, 17
Almond, 74, 80, 182
Almond moth, 181-182
Amyelois transitella (Walker), 181
Analysis
    logit, 37-41, 44-45, 51-56
    multivariate, 19
    principal components (PCA), 19-21
    probit, 6, 34, 37-41, 44, 48-51, 53-60, 62-63, 71, 90
Anarsia lineatella Zeller, 81-82
Anastrepha, 74, 121-122, 137-138, 157
Anastrepha fraterculus (Wiedemann), 139, 222
Anastrepha grandis (Macquart), 222
Anastrepha ludens (Loew), 23, 78, 80, 84, 105-106, 120-122, 133, 139, 150-151, 154, 213, 215-217, 222, 253
Anastrepha obliqua (Macquart), 79-80, 120-121, 137, 139, 151, 155, 222, 253-254
Anastrepha suspensa (Loew), 8, 78, 83-84, 105-107, 121, 125-126, 137-142, 151, 154-155, 174, 182-183, 204-206, 214, 216-220, 227, 239-240, 243, 250-251, 254-255, 260-261
Angoumois grain moth, 168, 182
Anthracnose, 91, 134, 136-137, 253, 256-257, 263, 267
Apple, 73, 79, 81-82, 84, 106, 108, 110, 121, 123-126, 150, 171, 174-175, 177-180, 184, 222, 230-232, 240, 250-251, 261, 269
Apple maggot, 78-79, 81-82, 121, 174, 204
Apple rust mite, 179
Apricot, 81-82, 109-110, 121, 123-124, 138, 250
Araecerus fasciculatus (De Geer), 78
Army cutworm, 18
Asparagus, 74, 84, 108, 171, 180, 184
Asparagus aphid, 108
Asynonychus godmani Crotch, 83, 97, 107, 140
Australian spider beetle, 17-18
Avocado, 73, 81, 84, 105-106, 110, 150, 152-153, 206, 240, 250-251, 254-255, 259, 269
    'Sharwil', 206, 229, 233
Azinphos-methyl, 97

Bactrocera cucumis (French), 92, 154
Bactrocera cucurbitae (Coquillett), 23, 73, 77, 83-84, 93-94, 101, 105-106, 120, 134, 137-138, 151-155, 202-206, 232, 240-241, 251-252, 254, 256, 259
Bactrocera dorsalis (Hendel), 23, 53, 56-59, 73, 83-84, 91, 93, 101, 105-

106, 120, 126, 134, 136-138, 151-155, 167, 182, 201-206, 232, 240-242, 244, 250-252, 254, 256, 259, 261

Bactrocera jarvisi (Tryon), 105-106

Bactrocera latifrons (Hendel), 201-202

Bactrocera musae (Tryon), 91

Bactrocera tryoni (Froggatt), 22, 91-92, 102, 105-106, 121-122, 126, 150, 154, 175, 204-205, 240, 253, 261

Bactrocera xanthodes (Broun), 234

Bactrocera zonata (Saunders), 106

Banana, 91, 93, 106, 109-110, 134, 138, 202, 205, 252, 259, 269

Banana fruit fly, 91

Baris lepidii Germer, 75

Bean
  dry, 74
  green or snap, 110, 152

Bean pod borer, 75

Bean weevil, 107
  four spotted, 119

Benlate, 91

Benomyl, 261

Benzyl isothiocyanate, 135, 205, 229

Binary quantal response, 50-51

Blueberry, 78, 121, 171, 174-175

Blueberry maggot, 77-78, 121, 175

BMDP (statistical software), 40

Boophilus microplus (Canestrini), 97

Brachycorynella asparagi (Mordvilko), 108

Broccoli, 110

Brown headed leafroller, 176

Bulb mite, 108

Cabbage, 80

Cabbage looper, 181

Cadra cautella (Walker), 181-182

Cantaloupe, 82

Carambola, 79, 83, 105, 107, 109, 125-126, 134, 140, 154-155, 205, 251-252, 255-258, 261, 264-266, 269

Carbaryl, 97

Carbon tetrachloride, 251

Caribbean fruit fly, 8, 78, 83-84, 105-107, 121, 125-126, 137-142, 151, 154-155, 174, 182-183, 204-206, 214, 216-220, 227, 239-240, 243, 250, 254-255, 260-261

Carpophilus, 93

Cashew, 79

Cattle tick, 97

Cauliflower, 110

Celery, 80

Cephus cinctus Norton, 17

Ceratitis capitata (Wiedemann), 23, 53, 56-57, 59, 73-74, 76, 78, 83-84, 93, 101, 105, 108, 119-122, 134, 136-138, 150-152, 154-155, 167, 200, 202, 204-206, 222, 232, 240-242, 244, 252-254, 256, 259

Cesium, 101, 106, 259-260

Ceylon gooseberry, 80

Cherimoya, 79, 110

Cherry, 73, 78, 81-82, 84, 105, 107, 109-110, 121, 123, 138, 231-232, 250-251, 261, 269

Chestnut, 74, 77

Chilling injury, 126-127, 257-259

Chlorpyrifos, 97

Choristoneura fumiferana (Clemens), 27, 53, 56-57

Citrus, 74, 78-80, 83, 97, 120-121, 125, 134, 140, 150-151, 153-154, 204-206, 213, 215-217, 240, 250-251, 254-255, 260, 269

Citrus blackfly, 74

Citrus flower moth, 74

Clepsis spectrana (Treitschke), 108

Climatograph, 17-19

CLIMEX, 23, 26-27

Climograph (see climatograph)

Cobalt, 101-102, 104, 107, 259-260

Cochliomyia hominivorax (Coquerel), 26

Coconut, 77

Codling moth, 72-73, 78-79, 81-82, 84, 104, 108, 121, 125, 171, 174-177, 227, 229-234, 240, 242, 251, 261

Coffee bean weevil, 78

Cold storage, 107-108, 119-127, 174, 178, 180, 185, 243-244, 257-258, 261

Computer simulation, 12, 17

Confidence limits, 34, 36, 39, 43, 52, 59

Conotrachelus nenuphar (Herbst), 81-82, 121, 179

Controlled atmosphere, 171-188, 240

Corn, 75-78, 167

Corn earworm, 181

Crab apple, 81

Cranberry, 78

Cryptophlebia illepida (Butler), 74

Cryptophlebia leucotreta (Meyrick), 121-122

Cryptorhynchus mangiferae (F.), 102, 106-107, 136-138, 167, 200-201, 261

Ctenopseustis, 171

Ctenopseustis obliquana (Walker), 176

Cucumber, 82-83, 106, 110, 151-152, 251, 254, 259, 270

Cucumber fly, 92, 154

Cucumis, 205

Curculio, 74

Curculio caryae (Horn), 166, 179

Custard apple, 79

Cydia fabivora (Meyrick), 75

Cydia pomonella (L.), 72-73, 78-79, 81-82, 84, 104, 108, 121, 125, 171, 174-177, 227, 229-234, 240, 242, 251, 261

Cydia spendana (Hübner), 74

Cylas formicarius elegantulus (Summers), 178-179

Date, 79, 110, 171

Degree day, 13, 16, 27

Delia antiqua (Meigen), 79

DEVAR, 16

Developmental rate, 12-14, 16

Diazinon, 97

Dichasmimorpha longicaudata (Ashmead), 219

Dicofol, 93

Dielectric heating, 165-168

Dimethoate, 90-92, 94-95

Distribution
  Gompertz, 37-38, 40-41, 44
  logistic, 37-38, 41, 51
  lognormal, 37
  normal, 37-38, 41, 43-44, 51, 63
  Student's t, 41, 43

Dose/mortality relationship, 35-37, 39, 53-57, 70-71

Dried fruit beetles, 93, 182

Drosophila, 126

Drosophila melanogaster L., 182

Dry heat, 149

Dysmicoccus brevipes (Cockerell), 97

Ecoclimatic index (EI), 23-24, 26

Ecoclimatograph, 17-18

EDB (see ethylene dibromide)

Effective Stress Value (V), 24

Eggplant, 82, 106, 141, 152-154, 256, 270

EGRET (statistical software), 40-41

Electromagnetic energy, 165

Electromagnetic spectrum, 165

Electron beam accelerator, 101, 103, 111

Endosulfan, 93

Environmental Protection Agency (EPA), 67, 134-135, 217

Ephestia elutella (Hübner), 173

Epinotia aporema (Walsingham), 75

Epiphyas, 240

Epiphyas postvittana (Walker), 108, 155, 171, 176-177, 183

Error, statistical, 33

Ethrog, 74, 123-124, 240

Ethylene chlorobromide, 250-251

Ethylene dibromide (EDB), 67, 92-93, 121, 133-137, 153, 199, 215, 217, 240-241, 249

Ethylene oxide, 251

European cherry fruit fly, 107

European corn borer, 75

European fruit scale, 178

European red mite, 179

Euxoa auxiliaris (Grote), 18

Exosoma lusitanica (L.), 75

Experimental design, 34-35, 39, 47-48, 50-51, 55, 62, 185
Export Certification Manual (USDA-ARS-PPQ), 68-69, 77

Faba bean, 75
False codling moth, 121
Fecundity, 21-22, 173, 185
Fenthion, 91-92
Fiducial limits, 41
Fig, 80, 109-110, 250
Film wrap, 182-183, 230, 250, 258, 271
Forced hot air, 149, 155-159
Frankliniella occidentalis (Pergrande), 79, 181
Frankliniella pallida Uzel, 108
Fruit fly (see specific species)
Fuller rose beetle, 83, 97, 107, 140

Gamma rays, 101-102, 105, 110, 168, 243, 246, 259, 261
Garlic, 75
Gaylussaccia, 77, 81
Genstat (statistical software), 40
Geographic information system (GIS), 27
Gibberellic acid, 219
Ginger, 75
GIS (see Geographic information system)
GLIM (statistical software), 40-41
Glutathione, 172
Grain mite, 180
Granary weevil, 167, 173
Grape, 73, 82, 110, 121, 123-125, 151, 240, 250
Grapefruit, 8, 74, 84, 105-107, 110, 123-125, 134, 140, 151,153-155, 183, 205-206, 216-219, 222, 227, 240, 243, 250-253, 255-262, 266-268, 270
Graphognathus, 75
Graphognathus leucoloma (Boheman), 80
Grapholita molesta (Busck), 69, 77, 125, 138
Greedy scale, 178

Green bean, 110, 152
Green headed leafroller, 176-177
Green peach aphid, 108, 171, 180-181
Greenhouse thrips, 155
Grumichama, 80
Guava
    cattley, 217
    common, 80-81, 106, 110, 120, 121, 134, 141, 151, 217, 254, 270
Gypsy moth, 27

Halotydeus destructor (Tucker), 74
Hawthorne, 81, 121
Heat sterilization, 149
Helicoverpa zea (Boddie), 181
Heliothrips haemorrhoidalis (Bouché), 155
Hemiberlesia lataniae (Signoret), 73
Hemiberlesia rapax (Comstock), 178
Hessian fly, 240
High temperature forced air, 149, 155
Hoboken Methods Development Center, 68, 70
Honeydew melon, 222
Horseradish, 75
Hot air, 149, 155-159
Hot water immersion (dip), 93-94, 97, 133-142, 167, 229, 232-234, 240-241, 244, 252-254, 256, 258-259, 261, 269-271
Huckleberry, 81, 121
Hypera postica (Gyllenhal), 17

Imported cabbageworm, 77
Imported crucifer weevil, 75
Index
    bioclimatic, 21-22
    ecoclimatic (EI), 21, 23-24
    moisture, 24
    population growth (GI), 24
    yearly stress, 26
Indianmeal moth, 174, 182
Infrared, 101, 165
Innate capacity for increase (r_m), 21-22

Integrated pest management, 97, 227
Ionizing energy, 101-102
Irradiation, 101-111, 168, 240, 243, 259-261, 269-271
Isohyet, 22
Isopleth, 22
Iteratively reweighted least squares, 41-45

Japanese beetle, 133, 250

Kaufmann effect, 14, 16
Khapra beetle, 107, 181, 183
Kiwifruit, 73, 121, 123, 171, 178-179
Koa seedworm, 74
Kumquat, 110

Large milkweed bug, 167
Latania scale, 73
Leaf miner, 97
Lemon, 107, 110, 153, 204, 251, 255, 257, 270
Lentil, 74
Lesser grain borer, 181-182
Light brown apple moth, 108, 155, 171, 176-177, 183, 240
Lime, 110, 204
Limonoids, 204
Linear regression, 35-36, 41-42, 44
Lobesia botrana (Schiffermueller), 73
Logit, 37-41, 44-45, 51-56
Longan, 110
Longtailed mealybug, 155, 180
Loquat, 110, 124, 217
Lychee, 84, 110, 121, 152, 201, 259, 270
Lymantria dispar (L.), 27

Macademia, 74
Macrosiphum rosae L., 181
Malathion, 173, 215-216, 218-219
Mamey, 80
Mango, 53, 80, 84, 90-92, 105-107, 109-110, 120-121, 133-134, 136-139, 141-142, 151, 153-155, 167, 182, 200-201, 241, 250, 253-254,

256-264, 270-271
Mango weevil, 102, 106-107, 136-138, 167, 200-201, 261
Mangosteen, 126
Manual of Fumigation for Insect Control (FAO), 67
Maruca testulalis (Geyer), 75
Maximum pest limit (MPL), 48, 61-62, 90, 93, 97-98, 234
Maximum residue limit (MRL), 89-94, 98
Mayetiola destructor (Say), 240
McPhail trap, 213-214, 216-218
Mealybug, 171, 183
Mediterranean fruit fly, 23, 53, 56-57, 59, 73-74, 76, 78, 83-84, 93, 101, 105, 108, 119-122, 134, 136-138, 150-152, 154-155, 167, 200, 202, 204-206, 222, 232, 240-242, 244, 252-254, 256, 259
Meloidogyne, 149
Melon fly, 23, 73, 77, 83-84, 93-94, 101, 105-106, 120, 134, 137-138, 151-155, 202-206, 232, 240-241, 251-252, 254, 256, 259
Melon thrips, 141
Merchant grain beetle, 181
Methoprene, 91, 94
Methyl bromide, 67, 69-72, 140, 153, 171-172, 183, 233-234, 240, 242, 244-245, 250-252, 261
Methyl iodide, 250
Mexican fruit fly, 23, 78, 80, 84, 105-106, 120-122, 133, 139, 150-151, 154, 213, 215-217, 222, 253
Microwave, 101, 136, 165-168, 261
Modified atmosphere, 171-181, 240
Moist heat, 149
Moisture index (MI), 24
Mold mite, 108, 180
MRL (see Maximum residue limit)
Muskmelon, 110
Myzus persicae (Sulzer), 108, 171, 180-181

Navel orangeworm, 181
Nectarine, 69, 72, 77-79, 82, 84, 109-

110, 121, 123-125, 138, 156, 201, 229, 233-234, 250-252, 271
New Zealand flower thrips, 171, 180
New Zealand wheat bug, 171, 180
Nysius huttoni White, 171, 180

Okra, 75, 77
Olive, 79, 110
Oncopeltus faciatus (Dallas), 167
Onion, 75, 79, 109
Onion maggot, 79
Opuntia fruit, 73
Orange, 8, 74, 84, 105-106, 109-110, 121, 123-125, 151, 153, 217-218, 222, 251, 260, 271
navel, 153, 251, 255
'Valencia', 153, 255, 259-260
Organoleptic evaluation, 264, 266, 268
Oriental fruit fly, 23, 53, 56-59, 73, 83-84, 91, 93, 101, 105-106, 120, 126, 134, 136-138, 151-155, 167, 182, 201-206, 232, 240-242, 244, 250-252, 254, 256, 259, 261
Oriental fruit moth, 69, 77, 125, 138
Oryzaephilus mercator (Fauvel), 181
Oryzaephilus surinamensis (L.), 182
Ostrinia nubilalis (Hübner), 75

Panonychus ulmi (Koch), 179
Papaya, 84, 93, 97, 101, 105-106, 109-110, 133-136, 152, 154-155, 167-168, 182, 202, 205, 229-232, 240-241, 244, 252, 261, 271
'Solo', 78, 134, 254, 256, 259
Passionfruit, 110, 124
Peach, 77-79, 81-82, 84, 105, 109-110, 120-121, 123-125, 138, 201, 222, 250, 252, 271
Peach twig borer, 81-82
Pear, 73, 81-82, 84, 110, 121, 123-125, 174, 240, 250
Pecan weevil, 166, 179
Pectinophora gossypiella (Saunders), 75
Pepper (bell or green), 84, 106, 110, 152-154, 256, 271

Persimmon, 79-80, 124, 250
Japanese, 155, 183
Pest free area, 8, 39, 94, 203, 213-222, 225, 233, 239
Phenotemperature nomogram, 14-15
Phosphine, 183, 251
Phthorimaea operculella (Zeller), 75, 84
Phyllocnistis citrella Stainton, 97
Pieris rapae (L.), 77
Pineapple, 74, 77, 79, 84, 93, 97, 110, 154
'Smooth Cayenne', 202-204
Pineapple mealybug, 97
Pink bollworm, 75
Pitanga, 80
Pitting, 140, 251-252, 254-256, 258, 262-263, 265-266
Planotortrix excessana (Walker), 176
Plodia interpunctella (Hübner), 174, 182
Plum, 77, 81-82, 84, 105, 109, 110, 121, 123-124, 174, 201, 250, 252, 271
Plum curculio, 81-82, 121, 179
Pomegranate, 78-79, 81, 123-124
Popillia japonica Newman, 133, 250
Population growth index (GI), 24
Potato, 75, 84, 109
Potato tuberworm, 75, 84
Prays citri Miller, 74
Principal components analysis (PCA), 19-21
Probit, 6, 34, 37-41, 44, 48-51, 53-60, 62-63, 71, 90, 201
Protein hydrolyzate, 218-219
Pseudococcus, 171
Pseudococcus longispinus Targioni-Tozetti, 155, 180
Ptinus ocellus Brown, 17-18
Pumpkin, 75, 106

Quadraspidiotus ostreaeformis (Curtis), 178
Quadraspidiotus perniciosus (Comstock), 108, 178
Quantal response bioassay, 50-51

Queensland fruit fly, 22, 91-92, 102, 105-106, 121-122, 126, 150, 154, 175, 204-205, 240, 253, 261
Quick freezing, 126
Quince, 77, 80-82, 121, 123

Radar, 101
Radio frequency, 101, 165-168
Radio waves, 101, 165
Radish, 80
Rambutan, 110
Raspberry, 110
Rate summation, 14, 16
Red flour beetle, 173, 182
Red imported fire ant, 19-21
Red-legged earth mite, 74
Regression, 33-44, 48, 50, 56, 242
Residual (statistics), 54, 57-58
Resistance
    to controlled atmosphere, 173, 185
    to fumigants, 183
    varietal, 8, 199-207, 218-219, 229
Rhagoletis cerasi L., 107
Rhagoletis completa Cresson, 83, 229, 233
Rhagoletis indifferens Curran, 19, 73, 105, 107
Rhagoletis mendax Curran, 77-78, 121, 175
Rhagoletis pomonella (Walsh), 78-79, 81-82, 121, 174, 204
Rhyizoglyphus echinopus (Fumouze & Robin), 108
Rhyzopertha dominica (F.), 181-182
Rice weevil, 107, 166-167
Rose aphid, 181
Rose apple, 217
Rotatable central composite design (RCCD), 35

Sample size, 33-34, 39, 51-53
San Jose scale, 108, 178
Sapodilla, 110
SAS (statistical software), 39-41, 54, 57, 71
Sawtooth grain beetle, 182

Screwworm fly, 26
Serial sampling design, 55
Shrinkwrap (see film wrap)
Sitophilus granarius (L.), 167, 173
Sitophilus oryzae (L.), 107, 166
Sitotroga cerealella (Olivier), 168, 182
Solanum fruit fly, 201-202
Solenopsus invicta Buren, 19-21
Sorghum, 77
Soursop, 110
Spider mite
    McDaniel, 179
    twospotted, 108, 156, 179
Spondias, 80
Spotted alfalfa aphid, 21-22
Spruce budworm, 27, 53, 56-57
SPSS (statistical software), 40
Squash, 110, 154, 175
Standard deviation, 33, 37, 40
Steam sterilization, 149
Stem-end rot, 253, 257, 263, 265, 267
Strawberry, 77, 79, 83, 109-110, 171, 181, 202, 250
Sugar apple, 79
Sugar beet, 80
Supraoesophageal ganglion, 108-109
Surinam cherry, 80, 217
Survival, age specific, 21
Sweet potato, 76, 80, 149
Sweetpotato weevil, 178-179
Sweetsop, 79
Synergism, 173, 185, 242-245
Systems approach, 8-9, 135, 206, 225-235, 239

Tamarillo, 110
Tangelo, 110
Tangerine, 74, 110, 123-125, 222
Taro, 78
Test
    chi-square, 37, 54, 57
    confirmatory, 47-48, 59-60, 90
    goodness of fit, 54-55
Tetranychus mcdanieli McGregor, 179
Tetranychus urticae Koch, 108, 156, 179
Therioaphis maculata (Buckton), 21-22

Thiabendazole, 257
Thrips obscuratus (Crawford), 171, 180
Thrips palmi Karny, 141
Tobacco moth, 173
Tomato, 76, 80, 82, 84, 90-92, 94, 97, 106, 109-110, 152, 154, 204-205, 259, 271
Transformation of data, 38
Treatment Manual
   FAO Quarantine, 102
   International Plant Quarantine, 103
   USDA-APHIS-PPQ, 68-69, 121-123, 133, 137, 138, 154, 240
Tribolium castaneum (Herbst), 173, 182
Trichoplusia ni (Hübner), 181
Trogoderma granarium Everts, 107, 181, 183
Trogoderma tarsalis (Melsh), 119
Turnip, 80
Tyrophagus putrescentiae (Schrank), 108, 180

Ultraviolet, 101

Vapor heat, 47, 149-156, 159, 254-256, 258-259, 269-271

Walnut, 72, 74, 78, 81, 108, 177, 181, 231, 234
Walnut husk fly, 83, 229, 233
Watermelon, 77, 231, 234
Weibull function, 56
Western cherry fruit fly, 19, 73, 105, 107
Western flower thrips, 79, 181
West Indian fruit fly, 79-80, 120-121, 137, 139, 151, 155, 222, 253-254
Wheat, 78, 166-167
Wheat stem sawfly, 17
Whitefringed beetle, 75, 80

X ray, 101-103, 111, 260

Yam, 76